Holistic
Agility

Rediscovering the Power and Meaning of Agile

Jim Lambert

Print ISBN-13: 978-0-692-14029-1
Ebook ISBN-13: 978-0-692-14030-7

Graphic Designs by Claire Hinkle Creates LLC

Contents

Preface

Snakes. They are symbolic of transformation. Now I'm not the biggest fan of snakes (picture Indiana Jones in the movie *Raiders of the Lost Ark* when he says, "Snakes – why did it have to be snakes?" – that pretty much describes me), but they are a recurring symbolic theme that represents very well the situation many of us repeatedly find ourselves in – the need for some type of transformation that will allow us to achieve new successful growth.

The reason for the transformation symbolism is that a snake's skin doesn't grow as the animal grows. The skin becomes a constraint and inhibitor that must be left behind to allow for continued growth but breaking free of it can be a challenging process. Not only is the process itself uncomfortable, another part of this transition period is the symptom of limited sight. The old skin of a snake has eye caps that become milky-white in preparation for shedding and may not come off cleanly with the overall skin. So, vision is limited before and during, and sometimes even *after* the transformational period. Even though sloughing off that old skin can be challenging, difficult and painful, when vision returns and the process is finally complete it becomes clear to see that a valuable and enabling outcome has been realized and achieved through that experience. The snake emerges transformed and ready to proceed with its next phase of life.

This is a book about achieving renewal through profound transformation. Agile transformation to be specific, but not with the same definition of what a significant portion of the business world currently believes "agile transformation" to be. Perhaps you yourself can identify with the snake analogy, having felt constraints in your organization that are holding you and your colleagues back. You may sense that something needs to change to allow everyone to become more successful. Perhaps you are currently in an environment that has very significant difficulties and challenges, and you are recognizing the lack of vision around you. That old skin is itching to be shed to allow for that sought-after growth and clear vision.

What motivated me to write this book was the positive feedback and results I have had sharing my knowledge and experiences (sometimes *painful* experiences) with many different organizations, technologies, teams and people. Given how this approach has resonated with them, I put together this compilation in the hopes that these perspectives and experiences can also be of value to you. From what I have seen, having at your fingertips a variety of ways to describe the intended outcomes becomes a method by which you can help many more people look at things from a different perspective, allowing everyone involved to come together and help their organization move forward to the next level.

That said, there is more than just business and consulting experiences contained in this book. All things that contain forms of "evolutionary transformation" are my passion. It is not what I do, it is who I am. I may be an Enterprise Agile Transformation Coach by day, but during my career I have also been drawn to study aspects of psychology, spirituality, martial arts, philosophy, Ashtanga and Kundalini yoga, history, mysticism, mythology, quantum theory, alchemy, meditation, religions both Eastern and Western, and characteristics of the subtle and energetic components of human existence. There are innumerable intertwining threads throughout all these topics that have transformational content embedded within. Every time I work with a company, a team or an individual, I attempt to bring the sum-total of that seemingly disparate set of knowledge to bear to help them gain new perspectives and understanding around the problems at hand, and to use it to then grow and transform into their next incarnation. The people I work with are simply people looking for knowledge and guidance. They want to be heard, understood, supported, and most of all, helped.

We have all been in a position where we were seeking some form of guidance and answers. That search frequently leads to passionately debated topics where there does not seem to be a clear winner. You are left with a situation where the choice is clearly yours alone to make. In a technology-oriented business world, there has been a history of these almost religious-like debates. Blu-Ray or DVD? PC or Mac? Microsoft or Java? Native or Web App? On or Off-Premise Cloud? Open Source or Commercial? I'm sure you have seen others that are relevant to your own context and experience. How to go about producing valuable solutions through technology can be a divisive thing. For many, the Agile approach to software development – the philosophy of "Agile" as marked with a capital "A" – has become one of those debates.

For those fundamentally polarized on the topic, Agile is either deemed a panacea, the solution to all your problems if you follow its scripted path; or "Agile is Dead", because it has become nothing more than a marketing ploy to sell you services and certifications that are a collective bastardization of the original movement. Zealotry reigns on both sides. Regardless of which side individuals gravitate towards, the business challenges still exist that led them to the debate in the first place.

This book is written for those wondering if Agile is something they should pursue to address their current business challenges regardless of their current technology. It is also for those who are considering taking another run at an Agile transformation because what they've experienced before didn't seem quite right. Holistic Agility is not intended to change the minds of those whom have firmly chosen a side in the debate, though it is my sincerest hope that for many it will at least spur deeper thought into why you believe what you believe. It is for those who want things to grow to be better and either don't know if Agile is for them or have attempted some form of Agile transformation and have not seen the desired results. Something was missing.

Holistic Agility as a viewpoint and approach to transformation is very much a "middle way" mindset. It does not consider Agile to be a silver bullet, nor is it "dead". Many organizations are still gravitating towards Agile, seeking it, pursuing it – and many are still struggling to find notable and material success. Through a collective set of disheartening experiences, Agile as a philosophy, as well as a pragmatic approach, has slipped into a state of decay. That may be the equivalent of "dead" for many, but a rebirth is possible. Agile has simply lost its meaning. That meaning can be rediscovered, the power of agility understood, and thereby the true benefits intended by the original movement can be unlocked. From what I have seen and experienced, those benefits can be, and have been, realized when Agile is approached from an organization-wide, holistic fashion. Hence the Holistic Agility book you have in your hands. I hope it provides some new perspectives that are of value to you as well as to the people you work with and for.

- Jim Lambert
Denver, Colorado, 20 May 2018
First Edition

The Decay of Agile

Agile has lost its meaning. It is as if the movement and philosophy has taken on some type of sickness. Collectively the business world has lost sight of the original intent of Agile and without that guiding light, the sickness has led to a weakened state of decay. Organizations and their people are not getting the vital benefits originally promised.

The extent to which the intent has been lost is readily seen in many of the organizations that have already sought out Agile as the sole answer to their problems. In these companies the business side of the organization believes the technology side is slow and expensive. They are falling short on innovative and high value solutions. Market share is in decline. Revenue goals are not being met. "We are losing" is a pervasive feeling.

From outside the organization comes a vague promise of hope. "Agile" is the "right" way to deliver solutions. "It's faster and cheaper" they are told. The terms and buzz words start to roll off the tongues of more and more business and industry leaders day after day. The pressure to conform builds. A perception rises that if you're not Agile, then you're behind the times. Well then, let's make our teams Agile!

In that last statement lies the root cause of many failed attempts and the resulting withering of the movement. By deciding that your focus will be targeted at a subset of your organization – in this case the "teams" – before you've even begun you have deemed those teams to be the symptom that must be treated in isolation with the prescribed solution of Agile. Using any treatment to directly manage symptoms will typically result in undesirable side effects. Just watch any pharmaceutical advertisement on television and you'll know what I'm referring to.

For any living organism, the environment in which it exists is as much a contributing factor to overall health as is the internal workings of the organism itself. Without being in an environment that provides the right type of nourishment and shelter, "dis-ease", decay and the process of dying are accelerated. A business is the equivalent of a living and breathing organism and needs to be cared for as such. Hence the need for a holistic approach to solving what is not a technology-only problem, it is an organizational problem.

For those that have tried, struggled and failed when attempting an Agile transformation, the true failure is they have not noticed and grasped the learning provided by the attempts that have already been made. By writing off an attempt as a failure it is not used as a learning experience that can be built upon to subsequently overcome the challenges at hand. For too many they have not used the painful experience for learning and growth. Instead they simply throw their hands up and say "Agile? Yeah, we tried that, but it didn't work here". No deeper thought or analysis is given.

Underneath the result of failure is the missed opportunity of stepping outside yourself to realize just how deeply old mindsets and behaviors pervade the organization. Take those deeply ingrained beliefs and behaviors, and pair them with a misinterpretation of the word "agile" in the first place, and that's when you get what leads to many companies not realizing true benefits - AINO. Agile in Name Only.

In the AINO scenario, the organization has incorporated new vernacular into their vocabulary – Backlogs, Stand Ups, Stories, Retrospectives, Pointing, Grooming, Demos – but when you look closer they are doing the same things they did previously, just with new words attached to those behaviors and activities. Yet the expectation is that the outcomes are going to be different. Of course, they are not.

Through the lack of organizational thinking and behavioral change, the lack of positive impact becomes apparent. For most of the people involved, their failed attempt has left a bitter taste in their mouth. The experience was not beneficial. It created chaos, confusion, more delays and overhead and an entire list of side effects. For many, the word Agile has become directly associated with those types of experiences, but the same business challenges remain.

Through the plethora of attempts that have struggled and failed, Agile as an industry term has now been saddled with that negative taint. Although having a negative connotation for many, in board rooms and during the hiring process, it has become almost required content that must be included in strategy decks and on resumes but is not being lived in daily practice. By many it is used simply to avoid the potential judgement of being labeled "NOT Agile", and they may try to claim some nominal form of success publicly while internally the true benefits were not even close to becoming manifest. Andy Hunt, one of the contributors to the Agile Manifesto, said it well recently when he said, "Agile now means, we do half of Scrum poorly and we use Jira". When you look at the typical challenges Agile implementations face, it is apparent that this is the case. People are going through the motions, but with little value being realized.

It has been clearly documented over the last several years as to why Agile Transformation efforts are not deemed successful. The top "challenges" are typically variants on these themes[1]:

- The company culture is at odds with Agile values
- There is a lack of experience with Agile methods
- There is a lack of management support
- There is a strong *overall* resistance to change
- There is a pervasiveness of "traditional development"

The notable takeaway here is that the challenges do not lie wholly on the shoulders of the technology and software development teams. It is not as simple as "the teams are slow and expensive". We are speaking about the *entire* organization and why it has been deemed that Agile didn't work in that environment. At the root, there remains a chasm between the business demands and the technology delivery capability. What often doesn't get directly questioned is *WHY* that environment is the way it is and *WHY* that chasm exists. If the business challenges existed prior to attempting Agile, and still existed *after* attempting to become Agile, then a more deeply probing question is...

What has made you and your organization believe you wanted to be Agile in the first place?

---

[1] VersionOne *11th Annual State of Agile Report.*
https://explore.versionone.com/state-of-agile/versionone-11th-annual-state-of-agile-report-2

Why Do You Want To Be Agile?

I've worked with many different organizations of all shapes and sizes. When an initiative is underway to "implement Agile", oftentimes it is difficult to get a succinct and clear answer to what may seem to be a simple question. Why do you want to be Agile?

For some there is an underlying perception that organizational credibility, and possibly the credibility of the individuals themselves, is at stake if they cannot say they are Agile. Many have claimed knowledge and experience in Agile and are now expected to apply it in their new job. Others have a perceived pressure that if they're not Agile then they're behind the times, perhaps lacking personal marketability. I've also seen companies proclaim to candidates during the hiring process they were already Agile and when those candidates came on board, especially those who had some prior knowledge, experience or success in Agile, they saw that this was clearly not the case. They felt like they got "the ol' bait-n-switch".

For companies where one or more of the senior leadership team members has jumped on the Agile bandwagon, the result is very often a mandate-driven approach to Agile transformation. It is a message delivered from on high to "Make our teams Agile", "We will become Agile", "Go be Agile", or some variant. This is how many leadership groups are entering their foray into the Agile world, saying and believing that it is "the solution". They make it "a top priority". What gets lost when that message is delivered to the broader organization is what underlying business problems they are trying to solve. Unfortunately, the resulting behavior is that the people in the organization feel like they have been judged and blamed for underperforming. The team members get some cursory or introductory training and theoretical information and are then expected to show near term improvements in their performance. They begin to go through the motions, mostly because the big boss said so, while not understanding how this new way of working is supposed to be applied in actual practice outside of the classroom where they were trained. While the teams are floundering, the rest of the organization continues to operate as they always have which causes even more problems for the teams. Soon afterward the grumbling and questioning on the floor begins, and then grows and festers as it continues to go unheard, overlooked and unaddressed. The transformation initiative turns out to be a failure without learning, and a stain on the term Agile itself.

For others closer to day-to-day delivery responsibilities, their answers to why they want to be Agile are often things like:

- "We want to get things done faster"
- "We want to be more predictable"
- "We want to be able to plan the work then work the plan"
- "We want consistency across our teams in the way they work"
- "We want to be more accurate with our estimates"
- "We want to meet our commitments to our stakeholders"
- "Our technology costs are too high. It is too expensive"
- "We are lacking in innovation and competitive edge"

So, what *is* Agile? Taking a moment to open up the dictionary and look up the word "agile", we see it has a two-part definition:

- *The ability to think and understand quickly*
- *The ability to move quickly and easily*

The reasons leaders and management say they are seeking to become Agile does not align with the definition of the word, and in fact often run counter to the intent of what agility is supposed to provide. Therefore, some deeper contemplation is required.

Many of the reasons given for wanting to become Agile are rooted in the shortcomings of being able to deliver in the current environment. If any of the above reasons are being communicated in your organization, it frequently points to an overall organizational environment that operates from a project-centric and plan-driven point of view.

Working in a project driven environment comes with some unspoken, and often overlooked, assumptions. One of the key assumptions being made is that the things being worked on in that environment are the things you *should* be working on. When the reasons given for wanting to be Agile are around trying to gain *some* level of certainty in getting *some* piece of work done in *some* certain time frame, it implies that whomever provided the idea for the project is effectively infallible. The normal behavior is that a project will begin, it will be worked through to completion and it will do exactly what it is expected to do. That is a combination of three inherent fallacies:

1. That before any work is started, you know exactly what to build and that it will have the intended impact on the business
2. You can predict the level of effort and calendar time that it will take to build this complex and nebulous solution
3. That you will realize the intended benefit when this nebulous something gets delivered through some fortune teller's vision of how the future will unfold.

That is intentionally facetious, but it highlights what are long held beliefs by many individuals and organizations that they can ideate projects, estimate them, fund them, staff them, and deliver them "on budget" and "on time". All of those things are illusions. They are guesses. They are not real. Reality only comes into play when you either *prove* that you have had the desired impact, or you *learn* that what you've done did NOT achieve the desired intent.

See if this sequence of events is familiar:

- The business wants to do some piece of work
- A request is made to estimate the level of effort to determine the time and resources required that are then used to calculate a cost to do that work
- The funding request is made for the estimated cost, and funding is approved
- A resourcing staffing plan is put together, making sure the resources aren't overallocated across multiple projects.
- A detailed, milestone driven plan is put together and stamped as "approved" or "baselined"
- Time and budget tracking and reporting is set up, etc.
- The project is kicked off

- And off we go for the next several months…. or more

Along the way reports are given as to whether the project is Green, Yellow, or Red. If things are tracking on-budget and on-time relative to the original estimates, then everything is Green. If things are looking like they will be late or cost more than originally estimated, then the color is changed, and tension levels rise.

Have you every truly considered what it is you are tracking? Your judgement of if this project is successful or not is based on these reports, so it would be best to understand if it is an accurate assessment, but consider this:

If the original estimates, budgets, and timelines were all guesses in the first place based on limited information and without knowledge of what will be uncovered and learned in the future, then all your judgements on what constitutes successful progress are only judging how accurate you are at predicting the future.

What you end up tracking is if you are spending the approved money at the planned rate. Nothing more. You could burn all the money and deliver something on time that provides no material value to the customer or the business. The project was Green all day every day, but it might have been a better business decision to never have spent the entire sum of money in the first place.

When you truly think about it, once you kick off a project there are not many opportunities to learn if you should modify what you are working on, or even completely stop the project. The estimated cost and timeline became more important than everything else. Everyone involved has lost sight of the business goal by becoming laser focused on project budgets and milestone dates. Although a significant amount of an organization's time and money is spent attempting to create some sense of certainty as to what will happen in the future, and then "reforecasting" or "re-baselining" when events unfold differently than foretold, it is typically not questioned as to why we perform those activities at all when they are consistently wasteful.

It is time to leave behind the false belief that you can get better at predicting the future through estimating how long it will take to get something done, with how many people and at what cost. Another way to understand the wasteful nature of these activities and behaviors would be, from a post-project perspective, to ask yourself if the outcome would have been the same, or better, if that work had never been done at all.

All of that said, I am not saying to abandon all planning activities, but rather to change what you are planning for.

> *"Plans are worthless, but planning is everything. There is a very great distinction because when you are planning for an emergency you must start with this one thing:* **the very definition of "emergency" is that it is unexpected, therefore it is not going to happen the way you are planning"**
> Dwight D. Eisenhower

In a software development sense, an emergency is not missing a planned milestone date, the emergency is realizing you are not going to have a positive effect on the customers and the business for the money that's been spent. Emergencies can be realizing you are about to miss the window of opportunity on gaining a competitive advantage or disrupting your industry. Focus your planning on how you will quickly *learn* if you have an emergency on your hands (i.e. you're not on track to meet your business objectives), and have a predictable response planned for how the organization will adjust gracefully when you realize that course correction is required.

THAT is the two-part definition of agility:

- *The ability to think and understand quickly*
- *The ability to move quickly and easily*

Being able to identify and adjust to an emergency is why you want to be Agile.

Redefining What It Is You Want

Being able to adjust and adapt to changing circumstances may be the reason to become Agile, but we cannot lose sight of the business objectives of being faster, cheaper and more predictable in day to day operations. Those are all valid concerns. So, if saying you want to be Agile for those reasons isn't necessarily accurate, what is it you want that will address those concerns?

Executive leadership is driven by financial principles. The goal is to have a profitable business through providing specific products and services to the market in a cost-effective manner. That foundation implies that all initiatives, projects, objectives, Key Performance Indicators (KPIs) and metrics for any business are rooted in terms of:

- Generating Revenue
- Cost Reduction
- Revenue Protection
- Cost Efficiencies

Rudimentary maybe, but executives and senior leadership want to know *every day* if the company is headed in the right direction to increase or maintain profitability. The longer the duration of your initiatives and projects, the longer you don't know if you are indeed on course. Extended duration of projects leads to increasing uncertainty. That uncertainty can paralyze you to the point of not being able to strike at opportunities that present themselves through an unexpected confluence of events. That means when something big does happen, you will not be agile in response to that moment.

What you really want is to know when that moment is upon you. You need to learn fast if you should keep doing what you're doing – *or not*. A project-oriented, plan-driven organization typically runs under the assumption that once a project is funded and kicked off that it will run its course in its entirety. While those projects are running through their entire extended duration, you wait. When you wait, you lose. So why is it that we expect the rest of the organization to march toward a fallacious project scope, budget and timeline when what we really need to know is what is going to move our numbers and make an impact NOW?

All the individuals doing all the work throughout the organization, both business and technology people, become laser focused on meeting project timelines, not on directly driving the numbers. There is no line of sight from what they are currently working on to *WHY* they are working on it. They cannot see the forest for the trees.

Each individual's drivers need to be aligned with the collective business drivers. The goal is to create a collective purpose common to everyone in the organization. This is the underlying business need for a Win-Win-Win environment that begs to be created. If the individuals on the teams know that together, what they do every day is valuable to themselves, the customers and the business, then they have a clear purpose that generates engagement and enthusiasm like little else can. Through that enthusiasm, the customers are receiving valuable products and services that make their lives better. When the customer's lives are made better, the results of that are seen in the bottom line of the business. The employees win, the customers win, and the business wins.

What you really want is to create and nourish this as a virtuous cycle. It is an upward spiral. In the same way Purgatory winds upward in a circle to Paradise in Dante's Divine Comedy, Holistic Agility is intended to pave the way so that the organization can walk that path. But there's a catch…

How do you know if each step you take is truly on an upward path when you can't see around the next corner? If you were truly walking that path, your brain would realize that the last step taken was slightly higher than the previous. You have in effect *learned* along the way that you are headed in the intended direction. You are learning with every step. Therein lies the true objective underlying the virtuous cycle we're trying to create:

- *How quickly can we LEARN that we are having a positive effect?*
- *How do we know that what we're doing RIGHT NOW is the right or best thing we could be doing for everyone involved?*

The redefinition of what we want is not to simply be Agile, it is to be a continuously learning and adaptive organization.

Agile Is An Enabler – Not A Solution

After attempting an Agile implementation, when a company continues to see that it is not delivering faster, or more predictably, or commitments continue to be missed, then it becomes easy for them to say, "Agile doesn't work here". The business problems are not any closer to being solved than before. Going into the transformation initiative they truly believed Agile was the end-game solution, when in fact they did not step back a little bit further to realize the larger system that they were attempting to influence.

If those involved in the transformation attempt had taken a broader perspective, it would have been realized that the type of Agile they were implementing was only targeted at a portion of the company. If they had continued down the path it would eventually have been seen that Agile was merely an enabler that needed to be subsequently leveraged to open the pathways to achieving true overarching business objectives across the larger organization. Agile in that targeted sense addresses just one cog in an organizational machine. It needs to be understood where Agile fits into the organization as a whole so that you can determine how other cogs will need to be tweaked for the entire machine to work effectively and efficiently. Beyond that, you also need to know what purpose that complete machine exists to fulfill in the first place. You need to know where you are headed with all of this.

At some point during your exploration of Agile you most likely have read or heard something similar to the phrase "Agile Is A Journey, Not A Destination". This phrase is true, but a journey in what direction? A nomad wandering the desert is on a journey as well but may not have any particular destination in mind. In a business sense, endlessly wandering the desert does not sound like a recipe for success. A destination needs to be set, even if you know that it will just be to visit for a short while before continuing your journey in pursuit of the next destination. For an organizational transformation to be successful you must know where you are going and what tools you have at your disposal to help you get there.

There is a book written by Paulo Coelho titled "The Pilgrimage". In that book the main character has been working for many years to earn his acceptance into an ancient mystical order. The culmination of his initiation ritual entails being awarded a sword. It is specifically *his* sword. When first attempting the initiation ritual, he is rejected from the order because it is realized that he has failed to learn the final and most important lesson – what to do with that sword once you have it. He then must go on a pilgrimage, a long journey on foot, to the destination of Santiago de Compostela in Spain to learn and understand how he is to use his sword to unlock true value and benefits to the world. If Agile is the equivalent of your sword that you expect to use to realize company-wide benefits, what exactly shall you do with it once you have it?

Determining how you will use Agile as the method to make your way to a target destination is a key contributor to enabling a successful transformation, but the initiative requires one other very important piece of input. To orient yourself toward your destination requires you know where you are *right now* relative to that destination. In ancient Greece, the term *"Gnothi Seauton"* was inscribed on the Temple of Apollo at Delphi. It translates to *"Know Thyself"*. It is important to truly know yourself as an individual and as an organization before orienting yourself toward a goal. Agile can only be a useful tool if it is used to help the right people at the right time. Your company's current state is going to determine if the timing is right. Many companies want to model themselves after the likes of Google, Spotify and Amazon. For most this does not take into account their own company's current state or true market and customer needs. Those example companies are already living and breathing Agile without giving it conscious thought, and probably don't use the term "Agile" at all because it is now inherent in their belief structures and behaviors. Being adaptive to current needs and opportunities is second nature. It is who they are, and they are continuously moving on to new ways of growing and improving.

For companies that have never made a concerted attempt at an Agile transformation, you may not be ready for using Agile as a tool. You need to understand where you currently are, and possibly need to do some preparatory work before attempting your change. You need to prepare the soil, plant some seeds and give them delicate care and feeding until the initial growth has taken hold and rapid growth can begin. You may need to garner support, identify opportunities for experimentation, or have a very painful project wrap up that can be used as a motivator to try something new. There will be more on the topic of how and where to begin in later chapters, but for now it is important to take time to know your people and know where you are in time, then decide if the contents of this book are valuable and applicable to your current situation and circumstance.

Now, if is apparent that Agile is your "sword" and you understand your current organizational state is ready to attempt an Agile change, the remaining piece is to clearly articulate a destination. As previously mentioned, "making our teams Agile" is not a valid destination unto itself. The destination you are heading toward is much more holistic – to be a company that provides highly valuable differentiated products to customers with ever increasing growth and profitability, built upon a foundation of being a unified organization that is constantly learning, adapting and growing. The goal is to move the entire company from a plan-driven mindset to an adaptive one.

What is needed to adapt to current circumstances is an underlying mindset in that often goes unmentioned during Agile transformations. That is the mindset of harnessing the power of "now". It is the difference between plan-driven and adaptive organizations. There is a confluence of innumerable events in any given moment. Chaos reigns. Instead of trying to understand cause and effect in hindsight and attempting to predict and control events in the future, powerful results can be achieved when you simply look at the environment as it stands right now and how you relate to it. The psychologist C.G. Jung noted in his foreword to an ancient Chinese manual of divination, called the I Ching, that "it is imperative to cast off the prejudices of the Western mind" and that "the Chinese mind, as I see it work in the I Ching, seems to be exclusively preoccupied with the chance aspect of events." "*We must admit that there is something to be said for the immense importance of chance. An incalculable amount of human effort is directed to combating and restricting the nuisance or danger represented by chance.*"

Companies constantly find themselves combating chance by creating and protecting detailed plans. The same mindset required to effectively comprehend the ancient I Ching is the same we need to achieve Holistic Agility. Look at the world around you right now, understand your current state in relation to that world, and decide how you can achieve supreme success by harmonizing with that world. You will need to reassess that relationship constantly. Predictability is fallacious. Therefore, you must admit that predictability is not a viable or reasonable objective, so you can then eliminate it and move on to something of true value.

Taking a Holistic Agility approach is oriented toward making the entire organization healthy, so it can interact efficiently and effectively with the outside environment. The goal is to create and maintain a harmonious symbiotic existence between the customers in the market and the products and services offered by your company. Understanding that Agile in isolation addresses only one component in an organization with many varied and component parts will allow a different mindset to begin to emerge. Agile is an enabler to achieving that broader organizational goal. The true benefits and power of Agile will be realized later, but achieving high performance first requires some training and conditioning in the basic methods. When teams are learning through repetition, and beginning to deliver on a consistent cadence, what many call the organizational "heartbeat" gets created. So when starting out with traditional Agile, it will provide you with that heartbeat, but what is a heartbeat worth if the body that contains that heart is not experiencing and interacting with the world?

The Sensory Nature of the Organization

Many organizations and departments have an omphaloskepsis habit. That's Greek for the practice of contemplating your own navel. They spend an inordinate amount of time complacently considering themselves and their internal workings at the expense of a wider outward facing view. A prime example of this is companies that have an annual budgeting process that is driven largely by identifying and estimating projects. With limited information and collaboration with the people that will actually be *doing* the work, a list of projects is planned, prioritized and funded with highly uncertain ROI numbers and dates. Since those funding investment numbers and dates become what is tracked against, they effectively become fixed targets in a de facto year-long plan. What's more, those projects often last more than a year before they are completed.

A company I worked with ran an analysis at one point and noted that their average project lasted 399 days, and 45% of those projects lasted more than 450 days. Another company had a large collection of projects with a budget of nearly $100 million planned out for the next 9 quarters, several of which were dependent on each other. Many companies, especially large ones, can be described in a similar fashion. There is an enormous amount of time and money spent planning and doing work on specific project ideas while the market and customers continue to evolve and change in the outside world. When you complete a multi-year project and finally pop your head up out of your hole to look around, will the market care? Will the customers still even be there?

Once a project is kicked off, there is little to no interaction between the people working on that project with the environment outside the company. For any living organism, interaction with its environment molds and adjusts behavior accordingly. It is crucial for survival. What direct touchpoints does your organization have with the outside world? How frequently do your project teams utilize those touchpoints for learning what is happening in the market and with the customers? Is the company constantly adjusting what is being worked on based on what you are seeing, hearing or feeling from those customers at any point in time? Continuous observation and interaction with the outside environment is essential for success.

There is a viewpoint in the study of neuroscience that the main function of consciousness – the reason that consciousness exists unto itself – is to combine all sensory inputs into a single reality for better planning and decision-making. Those sensory inputs are constant when you are awake. For a project-driven organization, the company's "senses" are in effect only used once every year or so to decide what the next set of projects should be. What happens less frequently is the company looking to see if the previous set of projects that were completed provided the actual value and return expected.

When the organization is focused on project execution and completion, there is no direct and constant interaction or feedback loop between the company and the outside world. If you placed yourself in a sensory deprivation tank for over a year, the world would not be the same when you came back out. You would spend time catching up on some of the most basic pieces of information as to what was happening in the world around you and how you fit in. Would you then choose to immediately go back into the tank for another year and do it all over again? Occasionally spending a bit of time in a sensory deprivation tank to get away from it all may sound heavenly to you as an individual, but from a business perspective success and survival are much more challenging in that scenario.

The next layer of that same neurological study really highlights why change is needed in a project-driven organization to achieve organizationally Holistic Agility. In the human brain, sensory input is being processed independently by different parts of the brain and then re-assembled to determine what is going on in the world. But while that re-assembly is going on, the world is changing. As a result, the so-called "current state" information is constantly out of date, and because each sense may interpret the same events slightly differently, the "current state" in its re-assembled form is potentially inconsistent with itself! This is also true in a business sense. Understand that your information is *always* out of date and that different parts of the organization may interpret the same information differently from their own perspective. The longer you wait to gather more information and put everything back together makes the overall picture *even more* out of date and, in effect, useless. You need to be constantly probing the environment to update what you think you know and adjust accordingly. As close as you can get to knowing what constitutes the "right now" will tell you the best course of action and your next step but be prepared to change course again the next time you assess the "right now".

The human senses only relay the here and now. That means that the past and future are illusions in that they don't exist outside of the mind. The past and future are completely internal constructs. So, the correlation when applying this concept to a business transformation is to minimize the amount of time and effort the company spends dwelling on the details of the long past and distant future. Doing so generates too much organizational "noise" which, when viewed through the analogy of navel gazing, draws away focus from the world outside, clouding the company's judgement and limiting their capability to react to current circumstances in the market and customer behavior. All of that "noise" and distraction begs to be quieted. The company needs to fully focus and concentrate on what is valuable and essential to success.

For anyone that has started a meditation practice, when you first begin you quickly realize just how much noise and chatter there is going on in your mind at any given time. It is almost impossible to concentrate. Because it is a *practice* it takes time and effort to be able to quiet your mind. Meditation is an integral part of yoga, which is more precisely the practice of unifying body, mind and spirit. For Holistic Agility purposes yoga is another good analogy because we want to bring the various parts of the organization together to form a unified whole in the same fashion. There is a need to begin a 'practice' that is intended to progressively quiet organizational noise, eliminating the multiple challenges for everyone's attention, and come to the point where you can fully concentrate on one thing for an extended time. In yoga it is taught that concentration leads to meditation which leads to Samadhi. Samadhi is "super consciousness" where union, a sense of oneness, is achieved. Once you have achieved this feeling of oneness on the inside, your interactions with the outside world are forever changed. Your capability to be fully present in the moment allows you to process the full depth, width and richness of the way things are and make clear decisions and plans based on the current circumstances. A holistically unified organization will be able to make those clear decisions and plans, and by effectively interacting with the environment in which they exist, adjust and adapt those decisions and plans continuously.

Being able to concentrate and become present in the moment has also been said to develop what is known as the Sixth Sense – intuition. Many people have a well-developed sense of intuition and have become comfortable listening to it. They "trust their gut". In a business setting, it's more likely that the executives and senior leaders will have the leeway to adjust course based on their intuition. As you go deeper in the organization, even though many will sense that what they are doing should change, they perceive they are not allowed to. They do what they are told, what the process states, what their boss assigned them, their tasks on the project, and so on. Their other senses have told and taught them that bad things will happen if they speak out or don't follow the rules. Your sensory perception creates your reality. Organizationally, with every group or person having a different perception of what their role is, there needs to be a change in what *everyone* believes to bring about unification.

Spencer Johnson, author of the book Who Moved My Cheese, wrote, "*When you change what you believe, you change what you do*". If you are hearing things like "this is how things are done around here", "the process dictates this" or "this is what we've always done", then you can see and hear that a new belief structure is required to empower the people on the other end of those comments. What Holistic Agility strives for is a new collective behavior and understanding of how to drive value. It is the "win-win-win" virtuous cycle mentioned in the chapter on redefining what it is you want. It is the "culture change" that so many companies set out to achieve but end up putting the cart before the horse in taking actions to "change culture" versus establishing a collective belief system that will allow them to change what they do and ultimately result in the positive and powerful culture they desire.

These first chapters have spent a lot of time describing the mindset of where many organizations have been and established that there are new possibilities when approaching things in a holistic fashion. A lot of the ideas presented so far may have seemed like they have nothing to do with Agile but they all provide different ways of thinking about how to approach transformational initiatives and communicate the need for change to others around you. Before beginning any transformational initiative, you must be able to describe why change is needed and gather some initial support before getting started. For anyone that has struggled to get support, perhaps these parallels, anecdotes and quotes will give you new ways to get others to finally understand what you've been trying to tell them all along.

If your intuition tells you that now is the right time to put something into practice in your organization, then the next logical question becomes where to start? If you go back to the "Let's Make Our Teams Agile" mantra, we can see why the software development teams are a common starting point for transformations, and it allows us to set the stage from a holistic perspective as well.

Development Teams – Agile & DevOps

Doing and Being are very different things. When the approach to an Agile transformation is targeted and mandated by leadership, the frequently made assumption is that if the software development teams change the way they work, if they "implement Agile", then the related business problems will be solved. The belief is that if the teams are trained and start to utilize the new process, then they will get things done faster, or be able to "plan the work and then work the plan", or whatever reasons were given as to why they should become Agile. The teams are told that this is the new direction the company is headed in, then they are given some introductory training and put back to work with the same pressures and demands that were there before *plus* the demands of showing they have adopted the new way of working. This normally leads to the result of teams *doing* Agile software development practices, but not actually *being* Agile. The overall environment is still quite rigid, and teams are rarely given the opportunity to identify and adapt to changing business needs quickly and gracefully. Even if the teams themselves begin to use some Agile methods and processes, by retaining the same team structures, as well as having the same pressures and processes *around* the teams, the teams are not in a position to *be* Agile.

An organizational focus on metrics and numbers is also treacherous territory for teams early in a transformation process. This is especially evident in companies where leadership and management tend to latch on to specific quantitative measurements to assess if they are Agile or not. Shining the spotlight on metrics like velocity numbers, or agile maturity scores, or number of features delivered are common pitfalls. The belief is that the numbers alone will tell them if they're getting better or worse. The focus on these numbers as a method of judging teams as "good" or "bad" leads to the behavior of teams doing things with the direct intent of managing the numbers so that the bright spotlight of organizational judgement shines somewhere else. They want to show outwardly that they are *doing* Agile as expected, but inwardly they are doing the same things they've always done just to get the job done in the only way they know how.

What frequently gets obscured and lost early in Agile transformations is the organization being able to determine if the changes being put in place at the team level are having any impact on the business problems that started them down this path in the first place. Stepping back and assessing the business impact may have to come at a later date, but that shouldn't keep you from pressing onward and upward using the software development teams as a starting point. Whether or not your organization has redefined why they want to be Agile, the transformation must start *somewhere* to get to the goal of truly *being* Agile.

> *"A journey of a thousand li begins beneath one's feet"*
> Lao Tzu

Most of us know the paraphrasing of that quote as "A journey of a thousand miles begins with a single step". Starting an Holistic Agility transformation is well suited to working with the software development teams as the first step. Small, controlled change from an organizational perspective is much more viable and achievable than trying to change all departments and initiatives concurrently. Not to mention that the team level is often where the broader organization frequently lays blame for what are ultimately systemic shortcomings. This is also an opportunity for the voice of the workers to finally be heard, since they probably have shrewd insight to the systemic issues that have existed up until this point and can now show where they can not only improve themselves, but where they will need help from the rest of the organization to be able to grow further.

When starting a transformation at the team level you can consider it the proverbial "drop in the lake" that will ripple out in all directions over time to eventually address any broader concerns, issues or shortcomings. This rippling effect can be visualized as a set of concentric circles that will eventually encompass the entire organization. For Holistic Agility purposes, it begins with a single team at the center:

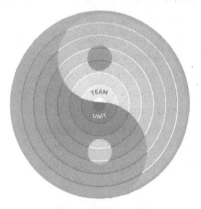

Pairing the rippling lake visualization with the analogy of a snake shedding its skin so that further growth can occur, the boundary lines between the circles depict the points at which other groups will have to change their behavior as well, shedding their own skin to enable new growth towards Holistic

Agility. As you ripple out from the team at the center, each additional role or department sits between two circles. Over the course of the next several chapters, this positioning will be used to highlight the most important touchpoints seen in their daily environment and responsibilities. It is how they fit in and play their part in a healthy and whole organization.

Starting at the team level is also a method that allows existing environmental inhibitors to be identified quickly. Getting a readout of the issues captured during a team's retrospectives or "lessons learned" sessions is a great way to realize that, although some changes at the team level are necessary and appropriate, others are related to the environment in which that team exists. Some things are in their control, but when listening to the teams what typically gets highlighted are issues that are systemic and institutional. For anyone that is driving the transformation initiative, or a change agent in general, they must be prepared to openly listen to and receive this type of feedback without defensiveness or justification, and then immediately take some form of action to improve. After a few retrospectives, if the team's calls for help are not being heard and broader organizational change is not occurring, these issues stop being talked about altogether while feelings of disempowerment and resignation set in. An unhappy team is an unproductive team that also becomes resistant to further attempts at change.

Often the team-centric "drop in the lake" analogy also describes the typical type of initial Agile training, consulting or coaching engagement. When the organization believes that the teams becoming Agile is the solution, consultants and trainers are brought in to "make our teams Agile". Since that is the business they are in, they will gladly accept the invitation to "make the teams Agile" through whatever products and services they offer. If they are experienced consultants that take a more holistic view, then they will know that a successful transformation will require much more than isolated change at the team level. Unfortunately, many trainers and consultants do not have that type of experience and are working under the same misguided assumptions that the organization itself is.

The first assumption is that the business issues and symptoms being experienced can be addressed with a surgical and prescriptive change. The other assumption, which is more difficult to address without significant resistance or churn, is that the way the teams are currently defined and structured will support an Agile way of doing software development. At the beginning of a transformation most team structures are not in line with the characteristics of a high performing Agile team. This can be clearly recognized when looking at the way software development teams receive feedback around the quality of the product they are developing and how quickly they are getting things done.

One of the reasons that the teams are a useful and powerful place to start your journey to Holistic Agility is that they sit at the center of a nested set of feedback loops. The feedback loops are the points at which you will learn if you are not getting the expected results and therefore need to do something different. Each time feedback is obtained it is an opportunity to take corrective action, meaning they are *all* opportunities to put some form of Agility into action. The generation of the feedback itself is through various forms of probing and testing the environment, typically consisting of one or more of the following:

- Unit testing
- Regression testing
- Integration/System Integration testing
- Functional/End-to-End testing
- Acceptance testing
- Product Demonstrations
- Production Readiness testing (performance and stability)
- Production Monitoring (testing to see if the live system is running as expected)
- A/B Testing
- Data & Analytics – (especially when testing to see if metrics and KPIs are being influenced as expected)
- Concept-to-Cash – testing to see if an effort had a positive ROI (return on investment)

By combining these test-oriented feedback loops with the organizational "lake" diagram shown earlier, it functions as a transformation roadmap for Holistic Agility. Starting with the team as the drop at the center and rippling out from there, you can progressively work with the various groups and teams on how to include these probing and testing methods as quality goals in their everyday tasks. For example, if you are currently in a waterfall development environment where coding is completed and then a new build is handed off to a group of manual QA testers, adding the activity of developers writing and executing a robust set of unit tests as part of their coding and build tasks might be a logical thing to do to help lessen the number of problems identified after that hand off point. The key at each feedback point is to understand *why* the particular testing activity has value. A test, as a feedback loop, is intended to learn something sooner rather than later. In this example, if unit tests are not performed, or not robust, then an issue with the way the software was written will not be identified until later, assuming it is found at all before customers report it. Waiting to learn is costly. There is plenty of industry data on the cost of fixing defects found in production versus those found earlier in the lifecycle, not to mention the cost of the impact of low quality work on customer perception. A foundational concept that applies to any Agile transformation is to leverage feedback loops to increase quality, eliminate wasteful rework, and reduce the associated cost and time spent on those activities. From a Holistic Agility perspective, that means we want to use feedback to learn fast.

As the waves ripple out through the concentric circles of testing feedback loops, we also begin to touch on the reason why the structure and composition of a team is critical to enabling quick learning. The more cross-functional a team is, the more types of testing they can perform without being dependent on another team. Dependencies equal wait time, which equals an increased risk of learning late, which is costly.

Cross-functionality of a team is a multi-faceted thing. In the most basic sense, a software development team is fully cross-functional when it has all the skills to design, build, and test a product feature across the entire technology stack, as well as deploy, operate and maintain that product. Many may interpret this as the definition of an Agile team, or a Team of Teams, or a Release Train, or a DevOps team, or any number of monikers based on the size of the organization, the software products it produces, and the technical maturity they have. Regardless of the label or goals that will be given to teams as part of a transformation, be it Agile or DevOps or both, it is oftentimes a far cry from where the teams are in their current state, and for most companies those functions are completed by such a large number of people that it is not viable to immediately form them into one tightly knit team. When deciding how to structure your teams, clearly defining what each team's purpose will be is the first step, followed by defining what skills each team will need to meet their purpose. Only then can you look at the people you have and map their current skill sets to those team needs and identify any gaps or skill set imbalances that exist across the broader organization.

This is where the varying set of methods and frameworks that most people equate with Agile and DevOps becomes highly relevant. Scrum, CI/CD Pipeline, Extreme Programming (XP) and Kanban are just a few examples of methods and frameworks to help teams "do" Agile and DevOps software development. As stated before, doing Agile development is a form of an enabler, but it is not Agile unto itself. Scrum is a common place to start for many project-driven organizations because it begins with breaking work down into smaller batches (via stories and sprints) and creates a continuous collaborative partnership between business and development people that usually was not there before. DevOps practices frequently have the goal of improving and increasing automation, creating a CI/CD pipeline, and extending the team's realm of responsibility through production delivery, deployments and operations. No matter the specific processes or frameworks chosen, they all can be foundational to, and enable Holistic Agility. For any of them to be truly valuable, you must know where your teams are today and if going to that level of capability and maturity is necessary, viable and appropriate based on market and customer needs. You can pick and choose from the various methods and tools available to make specific changes that are relevant to your current context. For example, if you believe continuous delivery and continuous deployment to be integral and necessary to what a DevOps team is meant to be, then you may start pushing that on your teams in the name of DevOps transformation and maturity – but if any of your teams are responsible for developing a mobile application, for example, they may not need to go to the level of continuous delivery and deployment in production because their customers may not want to be constantly updating their app. A by-the-book Agile or DevOps implementation may not apply in your current context, but perhaps certain pieces and parts of it do provide value to allow you to better meet your customer's current needs. It becomes a balancing act to keep the set of tools and processes as lightweight as possible but utilize just enough for the team to meet its intended purpose and provide value to their customers.

Enabling a team to effectively support customer needs also requires initial training and conditioning in whatever methods or frameworks are chosen. Getting the use of those methods to a state where they become second nature to the team and organically evolve from there can be challenging. The process that teams go through during this time is well represented by the concept of Shu-Ha-Ri. Shu-Ha-Ri is a martial arts concept that describes the stages of learning and mastery. It is used frequently when doing Agile coaching with development teams, but it can also apply to an entire organization. In "**Shu**" the student is learning traditional wisdom from a single master or teacher. The objective at this stage is to establish "muscle memory" by not deviating from what is taught until it becomes firmly engrained and instinctively executed. At the "**Ha**" stage, you may be learning from multiple masters, as well as begin to experiment and innovate around the methods that were learned during the "Shu" stage. This stage is a broadening of your viewpoint and knowledge base to include wider and more varied forms, options and possibilities. Finally, becoming a master yourself, you enter "**Ri**", where you may depart from standard methods and forms altogether, in effect becoming one with your surrounding environment and utilizing whatever method is appropriate in the moment. It is by progressing and evolving through these stages that survival is ensured. To be adequately prepared for potential challengers, you must constantly evolve your system, so foes and enemies cannot defeat you simply by your predictability. The evolution comes from taking something well known and adapting it to meet the needs of current circumstance, learning over time what works and what does not. The result is the best adaptation for what is needed right now.

Deciding on and training your teams in a core set of processes and tools is the same as the "Shu" stage of learning specific methods and engraining them in muscle memory through repetition. Once so engrained, the team can remain unified and strong when the environment shifts around them, generating the need to change course gracefully. By choosing methods, tools and frameworks, and getting them to become second nature through practice and repetition, you are preparing your teams to rise to the occasion when called upon to continue to perform well even when placed in difficult and challenging situations.

It is very important to understand and reiterate at this point that the power of Agility does not lie in whatever method, or methods, you choose to implement with your teams. The power is in the *people* on those teams and their capability to continue to perform effectively during times of change or stress. Therefore, another key characteristic of high performing Agile teams is that the longer they stay together, the better they perform. Dr. Bruce Tuckman introduced his "Forming, Storming, Norming, and Performing" model in 1965 which explains the phases a team goes through before becoming a high performing team. Being familiar with this model is key to enabling Holistic Agility, but it is also clear when organizations are not aware, or do not take into account, the basic human behavioral dynamics underlying team performance.

For example, in a project-driven organization it is a very common practice that "resources" are assigned to projects individually as each project comes through some form of intake and initiation process. More often than not, this is part of a waterfall approach to project execution, but there are also examples of spinning up "Agile" teams to deliver a project in a "water-scrum-fall" or some hybrid approach. Both of those delivery models do not address the true nature of team performance dynamics. Every time you put a different combination of people together, you send them back to the "Forming" stage, and thereby have an immediate and measurable reduction in productivity. *Every time*. Reductions in productivity are costly to the business. Yet many managers' jobs are to continuously reassign "resources" to projects, moving them around via spreadsheet allocations every time a new project or new need arises. Instead of taking this approach of bringing "resources" to the work, a shift must be made to *bring the work to the teams*. The longer a team exists, the better they will know each other which results in better productivity. Since projects have a shorter life span than the *products* that those projects create or enhance, the preference is to build and orient teams around the purpose of developing and maintaining a *product* over its inherently longer lifecycle.

There are added benefits of long-lived product development teams. Once a team enters the "norming" and "performing" stages of group development, they know each other's strengths and weaknesses well enough that, as a team, they become capable of operating in an independent and self-managed fashion. No longer is there the perceived need for the teams to be monitored and micro-managed by others in the organization because, through that independence and self-management, they become trusted units of the company. Long-lived teams are highly productive and have proven it by their results. From a holistic point of view, it is the same as not having to worry about what every organ or cell in your body is doing at any given moment. There is an infinite intelligence that is operating at that granular level without your intervention. Each piece knows what to do all on its own and continues to do so without you even thinking about it, but your body will also tell you when there is something wrong that requires intervention – you get sick. Let your teams stay together long enough to be able to do what they do in support of the whole without direct intervention and be ready to listen and take corrective action when they tell you that something is wrong.

A strong team dynamic also creates a sense of team ownership of deliverables and outcomes. A project-driven environment frequently facilitates and perpetuates the growth of individual egoistic goals, behaviors and drivers. That is the "I did my part exactly as I was told to do it, so it's not *my* fault we failed" mentality. Another characteristic of that individualistic mentality is when a project team is put on a "death march" to deliver a fixed scope on a fixed timeline and the due date is rapidly approaching. Heroes often emerge to save the day, and companies recognize and reward those individuals for their heroic efforts that allowed the project to be delivered on time. In a high performing team, the team succeeds or fails as a single unit, each doing their own part to get the shared deliverables across the finish line consistently and at a sustainable pace. When owning things in a true team fashion, they also soon realize that, as a unit of delivery capacity, they need to continuously identify how to get deliverables through their pipeline more quickly while meeting ever increasing quality standards. They can build on their successes and grow without burning themselves out. They can do more with less.

When shifting from a project-driven structure to one of evolving products, the removal of the project timelines can not only facilitate a sustainable pace for a software development team, it can also create a higher sense of urgency. If you are going to learn quickly what has positive impact on the customers and the business, then the team better be fast at getting things done, but getting things done *fast* does not mean working on multiple things at once.

One of the biggest misnomers in our current society is that multitasking gets more done in the same time frame.

Multitasking is a myth. This is a crucial theme to understand when working with any group of human beings and cannot be stressed enough as a foundational concept to realize the power of Holistic Agility. The human brain is single threaded. So-called "multitasking" is actually *task switching*. Task switching can be extremely detrimental. For example, texting while driving is not multitasking. You are making frequent switches between the task of driving and the task of texting. Each one is taking your focus for a brief amount of time, but not allowing you to truly focus on either task completely. The National Transportation Safety Board reported that texting while driving is the equivalent of driving with a blood alcohol level three times that of the legal limit. Similar research shows that up to 40% of productivity in a business setting is lost due to task-switching. That means it takes more time to complete the tasks you're switching between than if you were to focus on doing one task at a time in order. For software development teams, this relates to the concepts of reducing batch sizes and limiting work in progress (WIP) that most Agile, DevOps and Lean training courses will include in their content.

Limiting work in progress to a single item at any given time, and working that item through to completion, allows for doing the best possible job in the shortest time possible. It is the concept of *Presence* at work. Being fully present means having your focus, your attention, your thoughts and feelings all fixed on the task at hand. By breaking a team's work down into stories, features, pipeline items or some other small batch size of work that is not a full project, the team can become fully present and complete valuable items more frequently without context switching. Context switching is measurably wasteful in software development when having to do things like reconfiguring your development environment, checking code in and out, getting reacquainted with the code, finding your place to make the appropriate changes, etc. - but the costs of task switching can also be seen everywhere else in the workplace. Observe those around you and see if they are fully present and engaged in the task at hand, whether it is being in a meeting, listening during a phone call, writing code, or simply walking down the hall. Chances are you will see people trying to do other things at the same time. Checking their phone, working emails, instant messaging, or sitting with eyes glazed over as they think about something else. There is ample opportunity to apply the concepts of WIP limits and presence across the entire organization to enable higher productivity. Pay Attention. Be Here Now. Actively Listen. It benefits yourself and everyone else involved. Not only does it allow you to get the job done more quickly, one of the biggest benefits is that the quality of your work is also significantly improved by that all-consuming focus and limitation of work in progress.

But what is "Quality"? It is a relative and subjective term. Per the dictionary it is "the standard of something as measured against other things of a similar kind." We know it when we see it. For software development teams, it is one of the most relevant places to continuously improve. When a team creates an agreement as to the minimum set of tasks they will perform as part of the delivery of every single piece of work that comes their way, they are defining their current capabilities and limitations of creating quality deliverables. Most people know this type of agreement as a Definition of Done (DoD), where a team takes into consideration who they are at that point in time, including what skills they have and what they are capable of doing consistently to ensure their output is of the highest quality possible before it leaves their realm of ownership. It is also the place where they highlight their dependency on the capabilities of the broader organization. There are assumptions made as to what comes in via the work pipeline and backlog so that those tasks can be performed consistently by the team. Oftentimes these assumptions are explicitly stated when a team also creates a Definition of Ready (DoR) that establishes the state that individual backlog or pipeline items must obtain before they can be pulled, worked and delivered. Through these definitions the team has defined its boundaries and function relative to the rest of the organization. It establishes the innermost set of touchpoints and highlights where improvements can be made to enhance speed, quality and agility. Analyzing a team's Definition of Done (DoD) is one of the most effective ways to determine how far that team is from being able to be truly Agile.

When looking at the characteristics of high-performing Agile teams, there are a few key aspects that allow those teams to maximize their performance potential. Agile teams are the most effective when they are:

- Co-located to maximize face-to-face communication and minimize communication overhead and misunderstandings
- Cross-functional, possessing all the skills necessary to design, build, test, deploy, operate and maintain a software product across the entire technology stack
- Test-driven in their approach and automate tests as a normal course of business
- Able to directly access users and data for feedback

- Self-organizing, deciding who will do what and when based on their understanding of each other's skills, strengths and weaknesses
- Small (usually less than 10 people)

A team that has not been built this way, or has not yet developed these characteristics, will have a DoD that shows where they can further add to their skills and increase the efficacy and thoroughness of their feedback loops without unnecessary delays. For example, a team that has little to no testing automation may initially have a DoD containing items such as:

- Code complete
- Code Peer Reviewed
- QA Test Cases successfully passed
- Acceptance Criteria met

Whereas a team that is building more quality checks and automation into their everyday work may have a DoD containing more items such as:

- Code Complete
- Code Peer Reviewed
- 100% Code Coverage from Unit Tests – successfully passed as a "green build"
- Automated user/regression tests created and passed
- Acceptance Test Criteria translated into automated tests and successfully passed

The more cross-functional skills a team has, the more lengthy and explicit their DoD will tend to be. Many project-oriented managers get wary as this list grows, because in their date-driven mindset the risk has always been signing up to do too much in the time allotted. While the DoD shows what the team is doing to increase the quality of their deliverables, other characteristics of the team will allow them to do this larger set of tasks without adding to the duration of each deliverable. Those are the characteristics that minimize delays and wait time.

Many organizations have an offshore component to their teams. For many this is a remnant from when the hourly cost of an individual was the driving factor in utilizing offshore personnel for certain tasks. Over the years it has been proven that the hourly cost savings is lost when considering the delays and wait time related to the communication overhead required to get that work completed as expected. That is where small co-located teams bring true efficiency. Through face-to-face communication, especially for teams that have passed the "forming" and "storming" development stages, the capability to get immediate clarification can reduce *days* of wait time that would normally be spent sending new code off to be tested one day, answering clarification questions the next, and finally getting some type of results on the third day. A few minutes compared to days is typically *less expensive* than the assumed savings of lower hourly rates. The other speed factor comes directly from the energy and enthusiasm generated in that type of collaborative environment. If you have your designers, coders and testers all sitting around the same table, they are able to bounce things back and forth at a rapid pace:

"Hey, I just dropped a new build on the INT environment, can you check it out and make sure it runs with the API changes you just made? And while he's doing that we can be finishing up the automated user test

cases to have those run as soon as we prove the integration works. Oh, you found an issue? Let me fix that real quick and put a new build out there..."

Collective focus on a single unit of delivery, rapid fire communication, collaboration, issue resolution and quality checks. It all becomes a normal part of the team's everyday development activities. Everyone on the team leaves their title and their org chart at the door and works together to get each item to meet the Definition of Done. It is an amazing thing to behold when team ownership takes hold and things start clicking. The teams often are energized and seem like they are *happy* to be at work, which shows the true power of a high performing, co-located, cross-functional and self-organizing team. Happy employees stay with a company longer and push themselves and their teammates to continually get better.

When teams are able to pull from a backlog or pipeline of "ready to work" items, a team with these characteristics can rapidly deliver value. Their output is "potentially shippable" in the truest sense because they have done all the designing, building and testing themselves. While this is the scenario for teams to be striving for, when first starting out you will realize that there are organizational and environmental limitations that are inhibiting the team from reaching their true potential. The team will only be able to grow to the point where friction prevents their success from rippling out to the next level of the organization. That next level is the Product Management function that the team is dependent on to set their day-to-day direction and purpose. Product Management includes the people that define small batches of software development work and help get them to the "ready to work" state so the team can start. Product Management will also be the people who verify that all acceptance criteria have been met when the team is finished. We will turn our attention to the Product Management role shortly, but first, many larger organizations have multiple teams working on the same code base and the same products. If one of the characteristics of an Agile team is being small, then how do you address when you have many more people available with those skills that can increase your capacity and throughput? That is where scaling frameworks come into play.

Scaling Agile to Multiple Teams

When considering larger organizations, it is intuitively obvious to even the most casual observer that there can be virtual armies of people contributing the overall programs or portfolios of software development projects. Millions and millions of dollars are spent every year for all these people to produce the company's software products. Around all these contributors, and because of the level of investment, there is significant effort made to try and coordinate the individual projects as well as their dependencies across projects and teams. Certain projects may be dependent on the completion of a design or coding milestone from another project, or people with certain skills need to complete work on project A before being able to work on project B, etc.

When looking at the amount of time and effort spent detailing and tracking every interim milestone date and linking dependencies and "resources" across projects, it becomes clear that the overall plan that has been created is a very fragile house of cards. One misstep and the ripple effects through downstream projects and teams get immediately highlighted. What you often see is that a project, perhaps not due to be worked on or completed for many months, throws a flag by going into "Red" status because their prognosticated timelines won't be met due to delays in current projects. Frantic work begins on how to bring back in the timelines on current and future projects. Why is that? How can you possibly know that nine months from now you won't be able to complete a requirements document on a specific day because a current project's design document won't be done by next Thursday?

There is an enormous dependency model inherently built in to the way project teams are built and how they contribute to programs and portfolios of projects. Since you cannot predict at what point (nor how many times) you will learn that you didn't know exactly how everything would unfold when putting together a project plan, it is probably safe to assume that any detailed project plan will be wrong at some point. Therefore, there will *always* turn out to be dependency delays that come up as work gets completed, actual complexity of the work is revealed, and teams learn what is truly valuable through inspection of working software.

The constant discussion around delays in these types of environments often becomes too painful for leadership to continue to hear. They need a solution that will keep all of their portfolios and programs on track. This pain point is what leads many companies to pursue a scaled Agile implementation. They think Agile will make the teams get things done faster, and since they have a large number of contributors in the enterprise, then of course they must have a solution that scales to the enterprise!

Most organizations that attempt to implement a scaling framework believe that, since they have such a large number of teams and people, they absolutely must implement a framework to scale their Agile transformation across the entire company. For many this is a terribly misleading assumption. Not everyone with multiple software development teams *needs* to scale Agile. You need to know *when and how* to scale Agile. For organizations that do not yet have a strong foundation in Agile principles and also do not know the characteristics of an environment where scaling Agile is appropriate, they skip the step of re-organizing their teams to develop more key characteristics of Agile teams. Instead they place the immediate focus on the scaling framework terminology and processes. By failing to understand the core unit of a team and the added value of scaling true Agile teams, implementing some form of scaled Agile becomes another form of AINO (Agile In Name Only) that affects a much larger percentage of the organization. Some examples of symptoms that are often seen in struggling scaling implementations are:

- The team of teams not having the complete set of skills required to design, build and test a fully functioning end-to-end unit of working software, often waiting for a limited skill set that exists elsewhere in the organization to do their part (this is particularly true of companies with older, legacy components such as back-end mainframe systems)

- Running a Scrum of Scrums (SoS) that is nothing more than a Red/Yellow/Green status report meeting where the individuals in the room are not interested nor engaged until their name is called to provide their update, or a SoS that is more like a departmental staff meeting, informational update, etc.

- Framework trainers, consultants and proponents that insist that scaling needs to be applied because there are more than a handful of different teams, although each individual team is working on different and distinct products and are fully capable of delivering their product into user's hands without participation by other teams in the organization

- Established Project Management Offices (PMOs) championing a framework while not being able to articulate the value of the framework or how the new process and terminology is any different than what they've been doing previously

- "Big Room" or "Program Increment" planning sessions without software development team members in the room, where business and strategy teams attempt to put together a multi-year plan of projects and track it with a "program board"

- Large programs dividing the combined set of waterfall projects into arbitrary 2 or 3 week-long units and calling them "sprints" while doing "demos" of project documentation completed to date

- Starting to use the term "Done-Done" to describe things that must happen downstream from the development teams, typically additional testing and release readiness, often lasting many weeks between when a team claims "done" and the software is released to production as "done-done"

- The organization continuing to fund work as discrete projects, then building individual teams around the projects but grouping them into some form of category, although they work their backlogs independently and do not have combined efforts across multiple teams that contribute to shared deliverables

From the perspective of those experienced in Agile methodologies and scaling frameworks, it might be easy to criticize these folks and simply comment on how "wrong" they are and how "they just don't get it", but applying some form of empathy, understanding and emotional intelligence is much more useful to help them move toward a new understanding and to help them get where they are going. Most of these people are coming from long histories of doing things in a different way and have deeply ingrained belief systems and behaviors that they rely on when trying to get their jobs done. For scaling frameworks, especially for people that have lived a project-driven existence in their work life until this point, it's easy to gravitate toward the scaling frameworks because the amount of planning and process coordination is familiar and comfortable. Unfortunately, that results in many misguided people trying very hard to apply the methods but not being able to pull out the actual drivers behind *why* scaling is used. They simply don't know what they don't know.

If you are going to scale, it needs to be with the same purpose that an individual team has, being to learn fast. When you have multiple teams working on the same product, the key feedback loop for all of them is integrating their individual outputs into a cohesive, working unit of potentially shippable product. Integration needs to be done as frequently as possible to minimize the wait time before learning if the output of different teams does or does not work well when put together. Therefore, in the Holistic Agility roadmap, the circle that surrounds a single team testing their unit of software is multiple teams using integration testing as their main feedback loop.

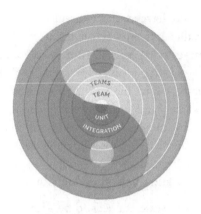

If the teams are working on separate code bases or products and there is no merging, integration or parallel development requiring system or end-to-end integration involved, then a scaling model will not add value. At their core, scaling frameworks are intended for use by software development teams and their surrounding Product teams to be able to collectively change course while remaining cohesive and aligned. Scaling frameworks also assume that you have built cross-functional feature teams around a product and no longer have dependent component teams that represent disparate skill sets. Just as a single cross-functional team aspires to have all the skills necessary to design, build, test, deploy, operate and maintain a software product, scaled environments seek teams of teams that include all the skills required to be fully cross-functional. If some skills are not on the team of teams, then you have remnants of a dependency model and are not scaling as effectively as you could be, and perhaps not at all.

There is an additional recommendation that oftentimes is not directly addressed in many implementations of a scaling framework. There is an opportunity for driving improvements in the quality of the product by having each individual team be responsible for both new development *and* maintenance and support work. Many companies select a scaling framework, build all their teams around supporting new feature development, and fill the backlogs with all net-new items. Off to the side there are separate teams responsible for coding fixes for defects found in production. For most, this is an unfortunate byproduct of having a budget that differentiates between Capital Expense and Operational Expense cost accounting, however it does not mean that the teams are required to be built as independent units to support the financial separation of tasks. What complicates things even more in that type of scenario is that there is a natural ebb and flow to new development and maintenance work. Having teams funded to do both new development and support work allows them to manage that pendulum of work through their backlogs without having excessive financial overhead and change requests. There will be more on the financial aspects of Holistic Agility addressed in later chapters, but the key point here relative to both team and scaling frameworks is that the quality of the product normally is better the first time through when a team knows that they are solely responsible for fixing anything that is found to be defective downstream. This is magnified in a scaling scenario because any production support work that is out of the ordinary will diminish the amount of time available for the new feature development that was planned and committed to by the team of teams. There is a lot to be said for teams that don't want to let their other team members and colleagues down or put them in a bad spot trying to explain why, when they said they could complete certain work, that something unexpected knocked the extended group off course.

When using a scaling model, frequently communicate with and coach the various teams participating in the implementation to avoid the pitfall of letting a scaling framework's well scripted, coordinated and interlinked set of processes distract from the true objectives. Remind them that the goal is to produce and deploy products and enhancements quickly to learn if they have positive value. When a course correction is needed in a larger development organization, a scaling framework can help the entire group pivot in a coordinated and agile fashion. When pursuing a scaling framework, it is very easy to over-engineer the solution. Compared to the discussion on a single team in the prior chapter, when multiple teams are working on the same code they are still working from a common backlog. To achieve their collective output, the integration testing adds the focus of producing an end-to-end tested valuable unit of potentially shippable working software. Whether using a scaling framework or not, there is a key Agile principle underlying the original Agile Manifesto that is very helpful to keep in mind when working with multiple teams and scaling frameworks:

Simplicity – the art of maximizing the work NOT done – is essential.

The more straight forward, unified and focused the team environment is, the easier it is to align the rest of the organization around that epicenter. This alignment typically starts with the teams' most direct touchpoint with the business, being product management.

Product Management

Establishing a true collaborative partnership between business and technology people is easier said than done. One of the 12 Principles behind the Agile Manifesto is "Business people and developers must work together daily throughout the project", but "working together" for one person may mean something completely different to another. When rippling out from the software development teams, the first "business people" we come across perform the role of Product Management. For many, a clear definition of what Product Management is, and is not, can be helpful in orienting Product Management with Agile software development teams. The definition of Product Management on Wikipedia states:

> *"While involved with the entire product lifecycle, the product management's main focus is on driving new product development. Superior and differentiated new products – ones that deliver unique benefits and superior value to the customer – are the number one driver of success and product profitability."*

In a project driven environment, the people labeled as Product Managers (or Product Owners, Capabilities Leads, etc.) are not doing what are normally defined as Product Management functions because they are not focused on identifying what constitutes "unique benefits" and "superior value" to their existing and potentially new customers. Within the confines of the project culture, the role of Product frequently turns into a requirements decomposition, prioritization and work sequencing function, if it formally exists at all. It ends up being somewhat of a half-business-analyst-half-project-manager role. This is what many people often mistakenly equate to the role of the Product Owner on a Scrum team. When a person that expects to be in a true Product Management role is put into that type of environment their innovative and analytical skills are curbed and stunted. They quickly start to feel disempowered, and that their voice, creativity and ideas are not being heard. They are wholly constrained by the approved projects they have been handed, and then are placed between the software development teams and whomever defined and approved the projects to manage the expectations of both parties. When trapped in this one-way flow of projects, they are not allowed to influence the outcomes as a Product Manager should. The value those individuals bring to the organization ends up being restricted and diminished, which is a loss for the company at large.

For Product people in these types of situations there are certain tools that can be of value when attempting to influence the organization to move from being "project plan driven" to one of "adaptive product evolution". There is an opportunity to understand and highlight the types of cognitive bias that exist in a project plan-driven organizational environment. From a psychology perspective, two types of cognitive bias that are pervasive in project-driven organizations are "anchoring" and "overconfidence". Anchoring bias is a very common tendency of people to overly rely on the first piece of information offered when making decisions and subsequent judgements. An overconfidence bias has two key aspects being the illusion of control (the tendency of a person to believe they have control when in fact they have none) and planning fallacy (the tendency to underestimate how long it will take to get something done). Planning fallacy is the strongest when planning long and complicated tasks or projects. It is appropriate to note that software development falls into the category of complicated and complex work.

Anchoring and overconfidence biases can clearly be seen when examining scope-driven project planning and budget activities. The very first rough estimates of cost and timeline for each project become anchored in everyone's mind. Milestones of detailed project plans are often "backed into" from the original date given. Tracking of projects once they are underway results in the product of reports stating whether or not things are on track to meet those (underestimated) first pieces of information provided. Because of these cognitive biases, the technology and delivery teams responsible for getting that work done are now labeled as "slow and expensive" because they are being compared to poorly generated expectations.

To counteract those types of natural biases, shifting to an environment of continuously learning, and proving anchors and plans are false beliefs, becomes a vehicle for producing value early and often. It allows the true value of Product Management to emerge. The potential of shifting the environment in this way is deeply rooted in creating a different way of thinking for many people across the company.

> *"A new type of thinking is essential if mankind is to survive and move toward higher levels."*
> Albert Einstein

It becomes very important to find a way to get individuals to apply critical thinking to themselves and begin to question their beliefs. This is one of the most difficult aspects of any transformation initiative. The heuristic (the process or method used) to spinning up new projects is based on a set of beliefs that is rarely questioned. Simply by defining a project with a certain scope, cost and timeframe for completion implies it is known how to solve the problem before the project even begins. In your company, for any given project, is it widely known who came up with the idea of doing that piece of work in the first place? How is it that this particular person's idea warrants spending a large sum of the company's money for such an extended period of time?

Oftentimes it is the leadership team themselves that define and decide what projects will be worked next. Or "the business" will decide independently from "technology" in an order-taking type of environment where, instead of being a fully collaborative partnership, technology is there to build whatever it is that the business tells them to and within the timeframe allotted. As mentioned before, that is often also the basis around which many companies' annual budgeting processes revolve. "We plan to do *these* projects over the next year and spend *this* amount of money completing each one". There is an inherent bias, and perhaps arrogance, that those ideas are the *right* things to do and that's how and why they get funded. For some, simply suggesting they might be falling prey to forms of cognitive bias will cause a behavior change because they don't want to admit they are prone to natural human dispositions. For others, they may need to be presented with a fundamentally different alternative approach that attempts to avoid cognitive bias altogether.

For difficult problems, what is known as "memetic evolution" has proven that the best solutions emerge from trial and error. A problem with inherent complexity, such as that of software development, obtains enormous benefit from a memetic approach. If you create teams with the collective skills required, you can allow the combined Product and Development teams to learn how to best solve the problem through controlled and deliberate trial and error and experimentation. The approach assumes that you don't really know what "good" is until you see it in action.

Whether it is leadership or Product deciding what will be worked on, the attitude of the entire organization must become one that makes them prove something is the right idea by getting something out quickly and determining if the company should continue to spend money on it or not. Then prove it again. And again. Keep proving it continuously until you can't justify the cost any longer. Currently, if a project is funded and started, it nearly always goes until "completed", even after multiple re-planning and re-baselining activities as things go awry. But is the project itself still even worth doing at all? Will the ROI you put in your business case even be relevant any longer if that project takes over a year to complete? You don't know what is truly valuable until something gets used by your users and customers. That is when you know if it is the right idea. Get there fast.

There is another crucial aspect necessary for the shift to the memetic, evolutionary approach of product development. There needs to be an understanding that "perfect" is the enemy of "good". In order to prove what is being worked on is the right thing to be working on, the changes to the product need to be released quickly. Projects are prone to bloated scope initially, with scope "creep" later, and often have large valuable functions truncated and cut out of scope altogether at the end of a project because they will not be fully functional or tested by the planned due date. By taking up the mantra of "is this good enough to release" on each small unit of functionality, market and customer validation can occur sooner because releases can happen more frequently and with more valuable content. The product may not necessarily end up having all the bells and whistles that can be thought of, but more often than not those are not heavily used features and functions anyway and as such have a much lower ROI in and of themselves. When Product can frequently and repeatedly put their eyes and hands on the potentially shippable increments produced by the teams, it will be revealed and determined in that very moment that what has been produced is "good enough" to provide measurable value or learning when placed in the customers' hands. This is also the point at which the team can move on to the *next* most valuable item and get *that* one done quickly in a "good enough" fashion as well. Small batches and units of work combined with deeming the results "good enough" is another way to reduce the wait time to get to the next feedback loop of Acceptance Testing.

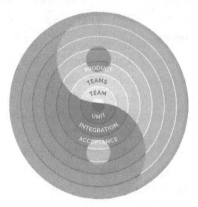

Acceptance testing, when paired with teams that consistently produce small batches of potentially shippable software, begins to build the content of the next product release. For a DevOps team with a CI/CD pipeline, each accepted unit may have the potential to be released immediately. For companies, products, markets or customers that require less frequent releases, each accepted unit is added to a release bundle and then the entire set is released at the next instance of the product's release cadence, such as an app update released through the app store. No matter how you get there, you are releasing as frequently as possible and appropriate so that you can learn if what you have done was valuable, if you should modify it and continue down this path or stop pursuing this path because the customer behavior and KPIs were not influenced as expected. You can then know that it is time to move on to the next idea. It is more important to release smaller batches frequently than to do a large, or "big bang" release that is often released later than originally planned. There is a very applicable quote that came out of Pixar films:

> *"We don't ever finish our films, we release them"*
> John Lasseter

Don't let perfect be the enemy of good. Release what you have built and learn if you should pivot or persevere, understanding that the cost of delay could be the difference between hitting a glorious window of opportunity or watching it close before you ever have the chance to get there.

Product Management, however, is very often part of a broader organization with many different demands and stakeholder groups. Therefore, part of being in a Product Management role is to incorporate these competing demands, rank them by a value-driven prioritization technique, and generate some type of roadmap or backlog. This roadmap creates a link to broader business strategies and, by being the tool used to prioritize and populate the team level backlog, creates line of sight for the team members so they know every day that what they are working on is contributing to the top business objectives. This is another place where, with the plethora of ideas and requirements coming from the entire organization, it would be easy to spend an inordinate amount of time prioritizing and detailing the entire list of everything the company could possibly want developed. Remember that what may be deemed priority and valuable now will absolutely change. Be mindful of using an approach to minimize and eliminate waste by using something like a rolling 6-month roadmap or feature list where the level of detail is becoming more complete through team backlog refinement work while leaving the later items sufficiently fuzzy until they work their way through a cadenced refinement process. If you spent the time detailing everything out now, when something else becomes priority tomorrow then that time and effort spent will have been wasted. There is a "just in time" and "just enough detail" mentality that can be used to getting a good balance of planning detail without being wasteful. Perhaps the outer reach of the roadmap timeframe is longer if you have an environment where you release software infrequently, say once a quarter, but no matter what the rolling time frame is, you can treat the entire list as a ranked "1-to-n" list of "next up" work and allow each individual item to be elaborated as it works its way to the top of the list when items before it get completed.

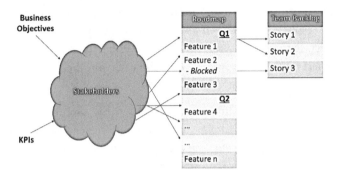

There is an associated diplomatic role for Product people that is part and parcel of managing a rolling roadmap or feature list. Product must be able to effectively manage stakeholder expectations. There is a great publicly available video by Henrik Kniberg called "Agile Product Management In A Nutshell" where he discusses stakeholder expectations management. The key concept is that stakeholders are usually asking "When will my stuff be done?" or "How much will be done by such-and-such a date?" or "Can my stuff be done by such-and-such a date?" Being able to answer any of these questions is dependent on enough team throughput or velocity history to create trend lines on a chart, as well as enough history and tribal knowledge of the team(s) to be able to do a rough estimation of level of effort necessary to complete items on the backlog. One of the most insightful statements in this video is around the third question, "Can my stuff be done by such-and-such a date?" Kniberg says that in Product management, although it may be tough, there is "No Lying". If your trend lines show that the work cannot be done in the time frame requested, you can say how much can be done by that date (through a reduction of scope) or the less desirable discussion of how much more time it will take to complete it all (resulting in a delayed learning and delivery of value). It is a very straight forward and clean approach to stakeholder expectation management, but there is another reason for discussing this here.

Although what Kniberg suggests is very useful when the contents of the roadmap or feature list are relatively short, it becomes a slippery slope in a larger organization with extensive demands and ideas that are captured in long feature lists and long-term roadmaps. To be able to answer any of the delivery questions for someone lower on the priority list, this approach would lead you to estimate much more of the entire product backlog and give a timeframe based on that information. This has a much higher risk of wasted effort, mismanaged stakeholder expectations and, at the core, implies there is some amount of stability and environmental control that would keep the roadmap and feature list stable until which time that stakeholder's work is next up. That is a level of stability and control which does not exist, so resist any pressures to go there. Do not answer the "when" question when a better response is to see if you can get that person to build a strong enough case that moves their work closer to the top of the list and becomes "now" instead.

To be an effective extension and partner of the development teams, product must gather all internal stakeholder requests, customer feedback and behavioral/usage data, as well as organizational objectives and KPIs, and distill all of that input into a prioritized and sequenced list of "next up" work. The prioritization function may become more of a facilitation function for Product Management in organizations with a large number of stakeholder groups, as the individual stakeholders must justify their relative position on the prioritized list amongst themselves through making a compelling business case argument as to why they should be ranked higher than another stakeholder that currently holds that position. This list becomes the input used by the combined product and development team to perform backlog refinement and ensure the team has a healthy pipeline of small, valuable units of work. Product Management getting work queued up for the team is the front end of a feedback loop. On the other end of this feedback loop is Product Management's acceptance of the completed units of work. By declaring acceptance, it has been learned that the software developed meets the business intent and can now be placed in the customer's hands to see if that value is ultimately realized. Once these practices have been achieved, the boundaries between product and software development teams start to become blurred. It is a daily partnership blending both business and technology points of view. One cannot survive without the other.

For Product Managers that have not come from a technical software development or operations background, it is very common for the roadmaps and backlogs to be slanted toward new features. The importance of this next statement cannot be overemphasized. **Technology absolutely MUST have an equal voice and representation during any prioritization discussion with stakeholders**. Any roadmap or backlog must be a combination of new development and system improvement efforts as both are required to meet business and customer needs and expectations. It may not sound as glamorous to do an operating system upgrade for example, but when it includes a fix for a critical security vulnerability that is a threat to your customers and business, then that work's position on the priority list or backlog can be communicated, defended and justified. Product Management, regardless of their background, needs to make a continuously conscious effort to have all stakeholder voices equally represented during prioritization discussions so that the maximum business value can be obtained through the multitude of forms that value may take. As the focal point of identifying how and when to address newly learned business needs and changes to business and system requirements, there will be a virtually constant negotiation of what bubbles up to the top of the list that is maintained by Product.

Finally, for Product Management to effectively arbitrate and prioritize the multitude of requests coming in, they must always have a touchstone, or Rosetta Stone, by which to translate and weigh the validity and potential value of any proposed work. Product Management must always know what the business objectives, strategy and related KPIs are that the company is trying to influence. By using this information to orient and drive value-based prioritization activities, both Product Management and the development teams know not only that they are always working on the most valuable things, they also know what to look for when software is released to see if the intended effects have been realized. Oftentimes this assists the teams in including new tracking, monitoring and analytics into the delivered software to observe and understand the effects once it goes live. Without this line of sight to business strategy, it becomes frighteningly easy to start to focus on output instead of outcomes, when what Product Management's main focus is intended to be is to deliver unique benefits and superior value to customers with the least amount of effort and investment.

Business Strategy & Stakeholders

Many companies have difficulty in identifying a distinct difference between their Product and Business strategies. This is often a symptom of the company being focused on project delivery or team *output* versus business *outcomes*. They end up tracking things like the number of features delivered but do not track if those features had any positive impact on the customers and the business. For Holistic Agility to be successful, a relentless pursuit of positive outcomes needs to be the collective purpose of everyone in the organization, but it is especially true of the combined Product Management and Business Strategy functions.

While Product Management is focused on *how* to influence customer behavior and KPIs, the Business Strategy function is *where* those KPIs are defined. Sometimes defining KPIs as part of a business strategy is the direct responsibility of the C-suite alone. Sometimes it is embedded in the Product Management, Marketing or Finance functions. Business Strategy can be a standalone department as well, especially in larger companies with more organizational complexity. Therefore, you may know this function by any number of names – Strategy, Product, The Senior Leadership Team, "The Business", etc. Regardless of the moniker used, this is where the decisions are made as to what work will be done and how that work will be funded.

In many of the large companies I've worked with, the Business Strategy group has been typically defined as the "owners" of delivering on the business goals and meeting KPIs or scorecard items. They are also allocated budget with which to initiate projects that they believe will achieve the goals. More often than not, this budget-driven project identification is represented as an annual occurrence, but when you look at the time and effort spent on these topics throughout the year, it is clear to see that it is nearly a continuous cycle. For example:

1. About 5 months before the end of the year the various business strategy groups prepare information regarding what projects they propose to work on in the next year, creating business cases with high level cost estimates

2. The next month, a lump sum budget number for capital investment is identified

3. For the next 3 months (or more) a negotiation battle goes on between the various groups, Finance and leadership to allocate funding to the proposed groups and projects, each claiming their portion of that budget until a "final" budget allocation for the next year is defined

4. Around the end of the year the PMO and all the Project Managers start to identify "carry over" projects that are in progress during the change in budget years, and work diligently to create tracking methods for how actuals will hit the current year's budget vs the coming years budget

5. Actual spend numbers begin to be analyzed during the second month of the year and start to tell the story of where projects are overspent or underspent relative to their allocated budget

6. About the fourth month of the year the corrective measure of a "6+6" budget begins to be planned to reallocate the budget based on actual spend as well as changing priorities

7. The "6+6" budget takes effect in the 7th month, which happens to be 5 months before the end of the

year, so the planning for the next year (see step 1) begins again

So not only is "annual budgeting" frequently a continuous process, it is also important to understand the negative results of this cycle. The first detrimental effect is that the entire organization becomes date driven. As soon as the projects get started, using the rough estimates of time and cost used to claim budget, the focus becomes on meeting milestone dates and not on delivering incrementally or iteratively to see what does or does not impact KPIs directly.

The second effect comes from the fact that the budget claiming project estimates were all created in isolation from each other. If there are more projects requiring a specific skill set to deliver than there are people available with those skills, then the "resource constrained" term starts to get thrown around. The question begins to be asked "can the business even spend what it asked for?" when it is a function of the number of technology people on staff and, as such, is completely out of the strategy team's control. The money was asked for before considering where capacity constraints might exist.

Thirdly, the uneven spending over the year across projects results in staffing volatility and "horse trading" behavior in the later months. Those who have projects that are spending money faster than the planned rate (those that are "burning hot") must let contractors go or move people on to expense work to get back within their budget allocation or seek to have money transferred from somewhere else in the budget in order to finish what they've started. Those that have available remaining budget become the targets of those who need additional funding to complete what they deem to be their own important work. He who has the money has the power, so all kinds of unfortunate behaviors manifest in the management and leadership teams in that type of environment. Either back room quid pro quo (scratch my back and I'll scratch yours) agreements are made, or remaining budget gets burned on lower priority work because those who have the budget know that if you don't use it, you will get less the following year. By now, with the laser focus on budget and timelines, everyone has completely lost sight of what is happening in the outside world. This type of environment rewards those who spend more by giving them more budget in subsequent years. When this type of negative behavior emerges, it often seems as if individuals in business strategy and senior leadership positions begin to define themselves by the size of their budget and that the year-over-year goal becomes to claim and retain the biggest budget possible as opposed to driving positive outcomes for the lowest possible amount of spend. It is inherently wasteful to spend all of that budget if the objective is to not be "shorted" the following year.

Timelines for strategy teams are another area of excessive wasted efforts. A 5-year strategic plan was the norm in business for quite some time. Many today still look at a two or three year set of horizons, but that is still fairly long term. Taking a quick look back to Dr. Jung's quote on trying to eliminate the unpredictable:

> *"An incalculable amount of human effort is directed to combating and restricting the nuisance or danger represented by chance"*

it speaks to how the enormous amount of effort put into detailed and long-term planning often ends up being an exercise in futility. While it is good to set multiple high-level horizons, transitioning to tactical delivery can be a 6-month rolling calendar at best. Anything beyond the next 6 months should be very rough and fuzzy, being much more thematic than defining the specific work to be done. For those items further out on the horizon, they should have little detailed analysis and planning performed around them because THINGS WILL CHANGE, and all that work will be wasted.

If you are part of a business strategy initiative, don't get overly attached to your strategic plan and convince yourself that this is how the future must unfold before you in order to achieve your goals. It takes much more time and effort to try and force things back into the plan versus seeing how things play out and adapting your plan to what happens along the way. Be prepared to throw your plan out the window on any given day. The environment you are in will tell you when it is time to do so. Whether that be through customer feedback, customer behavior, market volatility, competitive disruptors, new internal innovative ideas, whatever the case may be, do not let your commitment to adaptability be swayed by clutching to the old plan. From an efficiency perspective the goal is to minimize how much work ends up being wasted in those scenarios by recognizing that the longer the duration of the plan the more likely things are to change, which means the plan will be out of touch before you even begin. Let the present guide your decisions more than attempting to accurately predict the future by keeping the detailed tactical horizon as short as possible and the longer-term horizon fuzzy and malleable. Fortune tellers always say that the visions they see in their crystal ball are cloudy and hazy and are only *one possible future* anyway!

When you have a project driven budget and business strategy approach that is overly detailed and tactical, what gets lost is the necessary collaboration for learning, innovation and rapid success. With a full list of funded projects in hand to give to the technology teams, the business strategy team is in effect saying, "not only do we know what strategy we're setting, we're going to tell the rest of you exactly HOW we're going to get there!" What a Holistic Agility approach intends to achieve in the strategy realm is to move away from this type of "order taker" mentality where business strategy defines the work, gets the money then places orders to the technology teams. "Go build this for me, for this amount of money, in the timeframe I've set for you". In that type of environment, the projects flow down a one-way street, often with little to no collaboration with the people who will actually be doing the work. When switching from a project and plan driven environment to an adaptive one of evolving products with long-lived cross-functional teams, it becomes easier to identify what to STOP working on before you spend your entire budget.

Once you have some foundational transformation work completed in the software development and product management areas, the business strategy people can become much more intensely focused on the goals and KPIs. Business strategy people can utilize a management framework, such as a balanced scorecard, to define and track KPI's. The three key ingredients to making such a framework successful and valuable are:

- The choice of data to measure
- The setting of a reference value for the data

- The *ability of the observer to intervene with corrective actions* - meaning those who are monitoring the KPIs must have the wherewithal to initiate activities that will further influence the KPI results

As the "I" in "KPI", an indicator can only measure what has already happened. It cannot predict the future. Therefore, it is essential to differentiate between business goals, KPIs and target values. Business goals are the outcomes the company wants to achieve, which often come from top executives in the form of an over-arching vision for the company. An example might be something like expanding your market footprint by increasing the number of new customers registering to use your product. KPIs are the data points the business strategy group will want to measure to see if the work being done is having the intended influence and driving the company closer to the desired outcomes. This is where to set a reference value, such as mean time for a new customer to go through the registration process. Target values define the point where you want to see the KPIs trending toward during, and across the measurement periods. For example, the target is set for the mean time for users to complete the registration process to be reduced by 10 seconds from its current mean of 55 seconds because you have observed that at the 45 second point 60% of the users abandon the registration process. By setting goals, KPIs and target values that align with each other, and are things you can influence by how you create and modify your products, a scorecard framework becomes the center point of very valuable conversations on what to do next to continue to drive toward the company's strategic goals. It will also help to determine if you need to set new goals and KPIs altogether.

When determining what data to measure as KPIs, it is beneficial to define them in a "SMART" fashion. Make them Specific, Measurable, Achievable, Relevant and Time bound or phased. The "time bound" criteria is one of the most critical when looking at the goal of learning as quickly as possible. To create a balanced scorecard, the approach is to choose outward facing (i.e. customer and market) KPIs that have short lead times for gathering actual data and updating progress, and set internal KPIs for improvement initiatives that shorten lead times for other KPIs that may currently have longer wait times before learning if the results are trending toward meeting the business goals. That way all KPIs have an inherent focus on learning as quickly as possible. A balanced scorecard with this approach will tell you if the organization is healthy and whole by telling you if all KPIs are trending in the right direction.

The business strategy role creates the KPIs, some form of data and analytics work is used to update the scorecard, and executives and senior leaders use the output to assess the state of the company. Determining the health of the company is a question of whether or not they are successfully meeting customer and market needs, which is why it is also imperative to root as many KPIs as possible in the company's transcendent purpose and not use money-driven KPIs. Money-driven KPIs often set the stage for company and employee actions to become unmoored from the company's true purpose, leading to undesirable, detrimental and sometimes illegal behaviors. Looking at the interventions of the U.S Government into some of the largest financial organizations over the last several years gives some insight into exactly why this is a pitfall to avoid when defining your KPIs.

After defining a solid set of KPIs, business strategy people can work with product and technology to determine how and when to influence KPI numbers. The strategy is used as a key piece of input to help product prioritize and define roadmaps and backlogs that will keep the software development teams focused on the top value items. What business strategy can also do is keep a close eye on what is being built by the teams, and subsequently used by the customers to assess if the KPIs are the right KPIs and if they are achieving the intended outcomes. If there is any change from a strategy perspective, they can work with product management and the software development teams to reprioritize any next-up roadmap and backlog items as necessary. In early stages of a transformation, an effective way for business strategy to keep abreast of what is being built and used is through product and development teams performing recurring demonstrations of working software that has just been completed. Therefore, adding the "demonstration" feedback loop to our transformation roadmap adds business strategy to the picture:

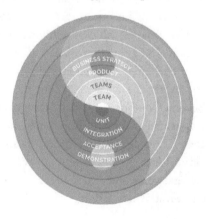

By seeing the next set of product functionality that has been developed, the business stakeholders may evolve or modify what they believe should be the team's focus over the next period of time. The demonstration frequency and content can vary depending on what type of development framework you are using. For a DevOps team using Continuous Delivery the demonstration could be what is running in production. For a team doing Scrum, it can be the increment of work completed at the end of a sprint. No matter where your teams are from a technology maturity perspective, it is always valuable to have stakeholders see the final product because it is often that it is when they *see it* that they realize that it is what they *don't want*. The more quickly you get that feedback the more you minimize wasted effort in the name of rework. This also means that you may or may not have additional production customer usage information at the time of the demonstration depending on if you are seeing working software pre-release versus post-release.

Decide whatever frequency and content works best in your current circumstance but know that the long-term goal is to increasingly build trust in the product and software development teams so that new functionality can be deployed without first doing a demonstration and review. This will allow reprioritization to evolve from a pre-release stakeholder review gate, to an analysis of KPIs compared to the realities of customer usage in production. If you are in a pre-release demonstration scenario, keep in mind that it is not merely a go/no-go decision. It is a feedback loop that allows for course correction through reprioritization of the roadmaps and backlogs. No matter where you are on that evolutionary path, product can facilitate the strategy and stakeholder analysis and discussion, and the combined team can update their roadmaps and backlogs according to what is deemed to be the right thing to do right now based on the current state of the software product.

As part of this iterative nature and evolution of the product roadmaps, feature lists and team backlogs, it becomes increasingly imperative that the product people understand the unified business strategy across all stakeholder groups, as a cohesive and synchronized set of KPIs and objectives, in order to be effective and successful. For the business strategy people themselves, to be able to create the strategy, KPIs and objectives, it is important to have a solid understanding of what can currently be accomplished on the existing technology platform(s) as opposed to what new technology or system architectural changes must occur before you can make that business strategy a reality. This is where it is imperative to have the business strategy thoroughly aligned with an enterprise architecture view of the world. In holistically unified organizations the two functions of business strategy and enterprise architecture have a symbiotic relationship that sets the stage for alignment of purpose across every individual in the company.

Enterprise Architecture

Build it and they will come. This is the unfortunate mindset of many enterprise architecture teams. While many software development teams themselves may have the responsibility to evolve a product's technical architecture, or create new solutions on new platforms, a formally separate and distinct enterprise architecture function typically exists in larger companies with more technical, architectural and system integration complexities. In these cases, the enterprise architecture group normally sets the vision and boundaries for the software development teams.

Stereotypically, executive leadership and senior management are frequently accused of chasing anything new and shiny like a magpie, but this is also true of many enterprise architecture groups. They are always on the lookout for anything that espouses new technology, new thinking, faster systems, cheaper development and maintenance, or reusable components. These are all good things in and of themselves, but not as much when approached with a project and plan-driven mentality. Many multi-million-dollar architectural modernization initiatives never meet their lofty goals and end up being minimally and sporadically used by the broader organization. Negative ROI on these types of architectural and infrastructure initiatives is all too common, if an ROI analysis is even done on them at all. There have been plenty of examples of these types of large-scale, re-architecting initiatives across a variety of industries. Web Application Servers, Web Services and EJBs (Enterprise Java Beans), Service Oriented Architecture (SOA), Enterprise Service Bus (ESB), Common Object Request Broker Architecture (CORBA), and

Client/Server were all technically progressive and alluring to enterprise architecture teams when they were new and "shiny", but developing an implementation and development strategy internally in isolation from the rest of the organization unmoored them from the goal of delivering business value rapidly.

As mentioned in the prior chapter, alignment between business strategy and enterprise architecture is a critical piece of achieving Holistic Agility, but frequently these two groups are for the most part operating independently and growing further apart. This is increasingly the case as the size of an organization increases. Independent enterprise architecture groups and initiatives have often been called an "Ivory Tower" as they work in isolation on the most aesthetically pleasing and technically current architecture they can model, but they are not working on developing those solutions in a memetic and evolutionary fashion through collaboration with the software development teams themselves. They are also not always engaged directly with the sensory parts of the organization that are identifying the current wants and needs of the customers and the market. The architectural designs may be theoretically pristine, but when there is an attempt to put them into practice through the software development teams, they fail to meet the needs and KPIs defined by the business strategy people.

As the enterprise architecture and business strategy functions grow further apart it is an all to frequent occurrence that they act as if they do not even speak the same language. The Ivory Tower produces a result more like The Tower of Babel described in the Old Testament Book of Genesis. The story of the Tower of Babel is about a unified people that originally spoke the same language. They undertake an ambitious initiative to build a grand and ornate tower that is intended to reach Heaven. During construction, their work is disrupted by an act of God that confounds their speech so they can no longer understand one another. The project is never completed, and they disperse across the globe, taking their own languages with them. That more accurately describes the relationship between enterprise architecture and business strategy in many companies. They simply no longer understand one another and are speaking two different languages, which drives an enormous wedge between the people on either side of the divide.

From a Holistic Agility perspective, enterprise architecture groups have two main responsibilities; to ensure that the current systems are stable and sound, and that the future systems will enable the realization of new business value and capabilities. Therefore, enterprise architecture is both a stakeholder and a business enabler wrapped into one. In many ways they function the same as the business strategy group in that they are identifying how to move the company into the future, especially when they are trying to address risks and weaknesses in the current architecture. Enterprise architects can justify a higher priority and position on various roadmaps and backlogs by being able to articulate the business value behind a technical initiative. That business value may be in terms of future savings such as reducing development or operational costs or addressing system stability or security risks. There are many ways to articulate the value of a technical initiative, but enterprise architecture must also remain one step ahead of business strategy in terms of delivery because extending the capabilities of the system architecture will also allow new business capabilities to be realized. This is another method of justifying a higher priority position because of the inherent dependency of a new business capability on the completion of an architectural improvement. So, it behooves the enterprise architecture group to be continuously synchronizing with the business strategy people in order to identify opportunities to do targeted, time bound proofs-of-concept that will provide the opportunity to learn if the architectural improvement produces the intended business value and paves the way for new capabilities to be developed and delivered to the customers.

Because of this interdependent relationship, business strategy and enterprise architecture must engage in an ongoing collaboration. Enterprise architecture must have a solid understanding of where the business wants to go so they can identify what is missing from the current architecture that would enable the realization of that strategy. At the same time, the business strategy people must have a thorough understanding of what is possible using the current system architecture as opposed to what new products, features and functionality will need to wait until the architecture is built out and proven to be ready for building new business capabilities. Therefore, in order to support and enable a successful Business Strategy, the feedback loop related to enterprise architecture is to prove that technical solutions are production ready.

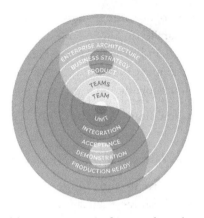

Enterprise architecture groups frequently spin up their own teams to do technical proofs-of-concept and initial development of new technologies and architectural components. The problem is that these efforts are often not aligned and embedded with the software development teams that will be responsible for developing, maintaining and integrating the overall production systems. After an "Ivory Tower" effort is completed by enterprise architecture teams, the expectation is that other teams and systems will integrate to what they've built AFTER they've built it. This is where the theory versus practice issues frequently arise, which is also why there needs to be a differentiation between enterprise architect and solution (or technical) architect roles.

Enterprise architects can define the vision, standards and guidelines that fully cross-functional software development teams can use as guidance, while solution architects can be embedded as part of the software development teams so that solutions can be developed and proven to be production ready through normal development efforts using a memetic and evolutionary approach. That allows new architectural components to be built and integrated as an inherent part of development in parallel to other related components of the end-to-end solution and overall systems and architecture. It is merely another example of the need for software development teams to integrate frequently and learn fast.

There is also an important Agile principle that is relevant to this approach:

"Continuous attention to technical excellence and good design enhances agility"

By aligning the current and future views of enterprise architecture with business strategy and allowing software development teams to put it all into practice, it becomes an important factor in avoiding an undue amount of technical debt. Teams are building, integrating and proving the production readiness of everything they build. No longer does the "technology" side of the house get put into a situation where they must "shut the front door", barring any new capability and feature work to enter, while working off technical debt, sometimes for a year or more. It becomes a balanced approach and partnership geared toward achieving technical excellence while concurrently meeting business needs.

What this also prevents is falling into the trap of doing technology only for technology's sake. That's why the business strategy feedback loop comes before enterprise architecture when striving for Holistic Agility. If there is not a need for a particular architectural capability to enable the delivery of specific business value, then there is no reason to spend the money on it. Not even to do a proof-of-concept. Architecture and development people can investigate those types of new technologies as a form of training to keep skills and analytic capabilities sharp or do it on their own time (as many do simply because of their passion for technology). There is value in that as well because they may be able to see opportunities in the future where that technology will become useful when a new idea for a capability emerges that could leverage that technology for the effective delivery of value to customers. These are yet more reasons to have a continuous circular relationship between enterprise architecture and business strategy where they synchronize and overlay their visions and high-level roadmaps to create unification in purpose and direction.

One item of note is that business process architecture teams are often a subgroup of enterprise architecture in large and complex organizations. They can be highly valuable in mapping business strategy to enterprise architectural roadmaps. Regardless of how the two roadmaps are mapped and aligned, the result will drive both the view of the future that senior leaders and executives will receive, and it will also define what type future state any operational and customer support teams should be preparing themselves for.

Customer Experience (CX) and Design

In companies that put significant time and attention into the look and feel of their products and services, there are often dedicated teams of Customer Experience (CX) or Design experts. While CX is frequently defined as encompassing all interactions that customers have with an entire brand, part of those customer interactions are through the company's software products and services. Therefore, you may also hear this specific role referred to as User Experience (UX), Human Factors, Experience Design, or simply "The Design Team". Regardless of the name used at a given company, the various types of teams in this space often wonder how they fit in when moving to Agile.

There is an added level of complexity when striving for Holistic Agility in an organization that is both project-driven and design-driven. Not only is there the pressure to deliver by a fixed end date, there is also a significant amount of time and effort spent focusing on how the final product's user interface will appear. Team members in this type of environment are constantly creating new designs, "mock-ups" and wireframes, often with a level of detail that goes all the way down to a pixel-perfect positioning and coloring scheme. The business strategy people and other stakeholders are often highly engaged in reviewing and providing feedback on these designs. This is something that is very easy for stakeholders to engage with because a design visual makes it seem real when simply looking at progress against project plan milestones does nothing to pique their interest. As human beings are inherently visual creatures, it is only natural that having a "picture" to look at seems much more valuable than hearing the project is in Red, Yellow or Green status. However, that picture they are looking at is counter-productive to the organization when trying to establish a sense of what is "real" right now. It is directly related to another key Agile principle:

"The primary measure of progress is working software"

That principle is most often referred to when discussing how hitting milestone dates on a project plan is not the best measurement of true progress, but it is also very much applicable in a design-driven environment. Stakeholders and designers working together often end up getting so consumed with the visual details that they unwittingly create for themselves an unrealistic sense that actual progress is being made on the solution because the picture is coming together beautifully. The stakeholders and business strategy people are collaborating over static pictures while not reviewing actual working software coming out of the software development teams. It is a natural place to gravitate towards, because the language and terminology of the designers is much more readily consumed and understood by the business stakeholders, but it puts CX and design people in a very precarious position. They are being asked to change the visual designs based on stakeholder feedback, but the stakeholders and designers very often are not working directly with the software development teams to know what each of those changes means in terms of technical viability, impact and complexity. When this happens it also misses an opportunity for innovation because the people on the software development teams are not presented with a problem to solve in a creative fashion while also being able to maintain technical excellence concurrently with good design. What these types of organizations have created is another form of the "order taker" mentality. "Build me something that looks like *this*. No, wait. *This*". The evolution of the software takes a back seat to the evolution of the visual design.

When incorporating CX and design teams into Holistic Agility, it is possible and viable to take the same type of approach as was described with enterprise architecture in the previous chapter. When CX or design is an independent function, they often have a split set of responsibilities. They must do forward-thinking designs of entire suites of products and services, such as rebranding the entire set of a company's apps and websites in a cohesive fashion, and they are also responsible for working with the individual software development teams to make sure that the actual solutions reflect those designs while they are being built. Typically, the more senior designers, such as creative directors, have the broad creative responsibilities that align with the branding objectives, while the more junior, or graphic designers are working on specific areas of a particular product. The senior designers act as the equivalent of an enterprise architect. Just as an enterprise architect creates a technology roadmap and vision, senior designers create a vision of the overall look and feel and promote consistency across all products and customer touch points through defining style guides and guidelines. The junior designers are similar to the solutions architects who are charged with implementing specific products and features. The junior designers use the style guides and guidelines created by the senior designers and work directly with software development teams to ensure that the working software aligns with the CX and design vision while the specifics emerge during iterative and incremental development. Therefore, it is also highly recommended that these designers are embedded within the software development teams in the same fashion as solution architects to create true cross-functional capability for the entire software development team.

Ops & Support with Data & Analytics

The people closest to the customers often have brilliant insight when it comes to what will make a difference in meeting customer needs, but those people are often left out of the equation when defining what is next up in the pipeline. With the advent of the DevOps movement, the value and the struggles of operational and support departments have been highlighted in many organizations. For many years in project-driven organizations the operations and support teams were being neglected while the rest of the business focused on getting the next projects out the door. Unless the system was "tipping over" in production, the attention was given elsewhere. Of course, systems having critical production operational issues was an all too frequent occurrence. The shortcomings of the rest of the organization became the daily burden of those people on the front lines of production operational and customer support.

Because of the quality issues that arise when rushing a project to meet a release date, the operations and support teams have learned that their interests need to be represented throughout the project life cycle lest they become responsible for more and more solutions that are difficult to support. These teams end up being split in two, always trying to know both what is happening in production and with the customers while at the same time keeping tabs on the details of what is coming through the delivery pipeline. Representatives from operations and support dedicate their time to sit in on project status meetings so they know what is being developed and make sure that the plan doesn't omit the things like support documentation and training that will be so critical to them when customers reach out for help and production issues occur. Like a salmon run, they work their way upstream through the software development life cycle to make sure they are part of the one-way flow from project initiation to production release.

Even when operations and support staff are at the table during project updates, they are frequently the first groups pressured to cut their tasks short or use incomplete information from development and testing teams so that the original date can be met. The main goal of meeting the release date remains all important yet the people who are expected to keep the system running and answer hard questions from customers have little to no idea what they're going to be running or what changes the customers will be seeing so they know how to guide them through the change. In the name of delivering on time, their project plan tasks get thrown out the window first with QA testing activities being close behind. They can see exactly what is coming, and it is not good.

There is another common symptom that arises from having a history of low quality solutions making it into the production environment. A "long tail" of activities arises between the completion of development and the release to production that becomes embedded in all future projects. It is often seen as part of a series of checkpoint "gates" that must be passed before a release is deemed ready for production. This often leads to the use of the term "Done-Done" where something may be considered "Done" when the development team completes what is included in their Definition of Done, but from a business perspective the work is not "Done-Done" until it is released. Some companies have created dedicated release management positions and entire departments just to facilitate the process from "Done" to "Done-Done". Code must be manually promoted from environment to environment, oftentimes having to remain there for a prescribed minimum amount of time before being able to move on to the next environment, eventually making it to production. I've worked with organizations where there was at least 6 weeks of lead time built in between "dev complete" and a release date – and that was the "happy" path! A lot can happen in 6 weeks. By requiring the software development teams to support and fix code in these interim environments while other groups put their eyes and hands on it for the first time, the development teams' capacity to deliver the next bundle of functionality becomes significantly reduced, and now *their* focus is split just like the operations and support people.

In technically complex organizations it is also typically deemed cost prohibitive for the progression of "lower" environments to mimic production, so the value-add of going through this process is dubious in many cases. For most, the entire process has been created as a result of the scars obtained in the past that turned into a fear-driven approach to delivering in the future. Time and effort is spent worrying about what *might* happen, and when nothing arises the code is promoted, and then things *happen anyway*. It isn't until code is released into production and used that any real learning happens, yet the "long tail" on the project has delayed that learning. One of the main objectives for Holistic Agility is to reduce the learning wait time.

Whether you have DevOps teams or separate operational teams, the operations and support functions of the organization are the place where a wealth of sensory data is entering the organization. Customer feedback in the form of usage analytics, production monitoring, support calls, online commentary and the like is constantly flowing in. All this data can be harnessed and leveraged. That is why the next feedback loop, collectively termed "Production Monitoring" here, is wrapped around the entire set of business and technology feedback loops:

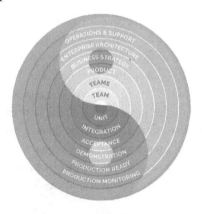

Operations and Support are direct customer touch points, running environments and helping customers with their problems. They are always watching and listening. They see what the customers are doing and hear how they feel about how well you are meeting their needs, but how often are these people involved in prioritization of what's next to be developed? Unfortunately, it is a rare occurrence, and when they do have the opportunity to voice their opinions they are often steamrolled by business strategy, enterprise architecture, product and leadership when the concerns being voiced are not in line with existing roadmaps and objectives. The company continues to be inward facing instead of continuously adapting to the environment they sit within that consists of the market and customers.

Customer support people carry an enormous amount of value potential both externally and internally. To tap into that potential, their voice and influence needs to first be unleashed. Externally they have the power to create highly loyal customers, but only when they are given the freedom to actively listen to, understand and help each customer they come in contact with. Striving for reduced call (or "handle") times and having call center scripts run counter to this objective. It is possible to let the support personnel be much more fluid in their jobs as they interact with the outside world and the customers. Having scripted responses does not help every customer in the same way, if at all, and rushing to get a customer off the phone definitely does not promote a customer-oriented mentality nor does it make those customers feel like they are genuinely cared about by the company. These are human-to-human interactions and need to be cultivated as such.

Support staff are the coaches and counselors of your customer base. When contacting you with a problem, each customer is going to be in a different place and have arrived there via a unique path. This is why scripted answers frequently don't work and why the support staff themselves need to be empowered but also have their own support mechanism. The goal is for the customer support function to operate using a key concept from the book 7 Habits of Highly Effective People:

"Seek first to understand, then to be understood"
Stephen Covey

Before being allowed to start coding at my first job, I was required to work in the call center to learn more about the software products we provided. We did not have scripts written that were to be read verbatim to a customer on the other end of the line, nor did we have a target by which we were to limit each call duration. I was able to take the time to understand the customer and their problem, often while they were under duress to get their own job done, before walking them through attempts at solutions or teaching them methods that they may not have known about. The company had also set up a mentoring program in the call center, so during calls that I didn't currently have the knowledge to solve, I could place the customer on hold and speak with a mentor to get guidance on how to resolve the issue. This provided rapid education and growth for me as a support technician. I also had many instances where customers were highly agitated, basically ready to throw their machines out the window, but I was able to take the time to calm them down, walk them through things, and truly help them. Several called the company later to say just how pleased they were with the type of service we were providing them.

People are looking for help, do what you can to empower your support staff to help them. When support people are able to use all the tools at their disposal to help each individual customer, both sides will feel the intrinsic benefits of the interaction. Beyond that, the transcendent purpose of the company will continue to be met through meeting customer needs.

From an operational standpoint, if the system is not up and running you not only aren't meeting customer needs, you are also not learning what's happening in the outside world. System health and customer behavior is often monitored and assessed from an operational standpoint using scripting, data and analytics. Not only can operations and support people use the info on the fly to address immediate issues, the information obtained can also be supplied to the rest of the organization as a form of feedback regarding overall health and effectiveness of the solutions and the organization.

As interactions with customers occur, the data and analytics gathered can be fed back to any of the other teams. For example, reporting actuals against KPI targets would be useful for strategy and leadership people while providing data on system health would be valuable input to enterprise architecture initiatives when determining where system improvements are most needed. That said, be wary of the pitfalls of using metrics, data and analytics as feedback loops. It is very easy to obsess over the amount of data that can be gathered, and it becomes easy to end up tracking more metrics than can be used efficiently and effectively. Companies that have extensive lists of metrics can fall into the pattern of not being able to see the forest for the trees.

To combat the potential for metrics overload, many companies have gone in the opposite direction and defined the One Metric That Matters (OMTM). They orient all initiatives to drive changes in that metric until a business inflection point is reached. Others utilize things like the balanced scorecard mentioned earlier in the chapter on Business Strategy & Stakeholders. Regardless of the number of things being tracked, the organization must constantly question if they continue to be the best metrics and KPIs in terms of signaling improved performance against the business goals and objectives. While they may be useful now, they will not always be the right things to measure in the future when circumstances change, so always question their continued relevance. Data and analytics can be the brain power of the organization, but only when you can monitor and act on what is relevant. By releasing new code into production, you are performing an experiment to see if your foray into that space influences the data. By shortening the time between releases and tracking the most relevant data, you can drive towards continuous processing of what is happening in the environment and adapting the company's response in the most beneficial manner. Each time this is done it is another opportunity to define what remains relevant and if any new approaches should be pursued. This continuous cycle enables the sensory nature of the organization to be focused on the present, then adapt and thrive accordingly.

Once you know what your business goals are, you need to know if you are making progress toward those goals. Good metrics will change the organization's behavior while bad metrics only contribute to information overload. Taking an approach of simplicity, using uncomplicated comparative metrics with rates or ratios, will assist in driving towards intended outcomes. Absolute numbers are rarely useful because they don't tell you if the trend is moving in a positive or negative direction, so behaviors will not change when provided with an update on that type of metric. For example, saying you have 96,000 new customers does not influence behavior, whereas converting that to a rate or ratio, such as noting a 10% increase of new customers over the prior month when the goal was 5%, or that you are less than 1/10 of the way to the goal of having your first million customers will spur more discussion on what to do next.

For companies that have large amounts of data and are performing "Big Data" analytics, it is also possible to uncover hidden patterns, unknown correlations, market trends, customer preferences and other useful information. If used wisely and thoughtfully, this type of data analytics can procedurally be treated as a stakeholder unto itself because the information identified can either be trended and tracked as KPIs or used as input to planning activities by generating new work items on roadmaps and backlogs. This may be more of an evolutionary goal, and unnecessary for many, but it speaks to limiting the number of things you are tracking at the beginning to understand if you are truly making progress toward the business goals. By starting small you can progressively determine just how deep into the analytics world you really need to go to drive success in the business outcomes. It will keep not only the operations and support teams focused, but the entire organization aligned to the pursuit of outcomes over output.

To remove the "salmon run" behavior of operations and support people mentioned earlier in this chapter, those groups can leverage communication mechanisms already put in place by other teams. One of the most valuable options available is attending demonstrations of working software as it will educate them on what is *actually* being delivered, not what *might* be delivered through a project. For larger organizations with infrequent releases due to the need for call center training or the like, perhaps the trainers are the demonstration attendees that will attend and then relay that information to actual support personnel through formal training prior to the launch date. For a more forward-looking view, operations and support teams should also have access and input to any roadmaps and backlogs being produced. The operations and support people have equal voices during a prioritization discussion in the same way that technology and architectural representatives do, being able to suggest new initiatives and provide business justification for higher prioritization rankings. They are stakeholders in the organization just like any other group.

Finally, the combined business strategy and architectural roadmap can provide the big picture view for anyone that might benefit from the high-level information of what is on the horizon. Through all of these communication mechanisms the people with the most direct customer interactions can be supplied with the information they need to effectively do their jobs, and even the topmost executives and senior leaders will have the opportunity hear their voice regarding what new work might provide added value to those customers.

Executives and Senior Leadership

All executives are not leaders, and all leaders are not executives. Leadership is not a function of organizational hierarchy. It is *who* you are no matter *where* you are.

From a typical roles and responsibilities perspective, the executive and senior leadership teams set the company strategy, make the biggest risk-versus-reward decisions and ensure that the company's operations align with and support the achievement of strategic goals they are setting. Although it is correct and proper for executives to be the people making these top-level decisions, for many it comes with a sense that they also direct everyone in the company regarding how they go about their work to meet the goals that have been set. This is not the definition of being an executive, it describes the difference between being a manager and a leader. Managers tell you *what* to do, while leaders succinctly identify the challenge to be overcome, create an environment geared for success, and support others in identifying how they will succeed collectively. For a true leader, the focus is on the people much more than it is on the work. The problem for many companies is that they have a plethora of managers and a dearth of leaders. It takes true leadership to identify a meaningful goal, articulate it to the entire organization, garner ever increasing support, and relentlessly drive the organization to that goal. Many so-called leaders delegate these responsibilities, allowing others to stumble along in an incohesive fashion, eventually allowing the journey to be abandoned.

When pursuing an Agile transformation, the term "command and control" is often used to describe the type of mentality that exists for the management teams in a project-driven organization. The intent of any Agile transformation effort is to change the management mentality to one of employee empowerment and decentralized decision making. Executives are often the most egregious violators of this intent, as they continue to seek a sense of certainty about future outcomes as opposed to embracing the uncertainty inherent in the environment. They are notorious for not educating themselves on the what the transformation entails for the various teams while continuing to exhibit old behaviors such as demanding exactly when projects will be complete, frequently reorganizing team structures, and focusing on increased output versus improved outcomes.

When a transformation initiative is working outward from the software development teams it is often the executives and senior leaders that unwittingly perpetuate the business problems that led to the desire to transform. While they use the current buzz words and expect the teams to change their behaviors to make their problems go away, they fail to take the courageous step of looking to transform themselves as well. They are not living what they are prescribing – not practicing what they are preaching. Holistic Agility cannot succeed without *everyone* coming together to create a collective set of shared beliefs and behaviors. Executives must be willing to look at themselves with just as much of a critical eye as they do with the teams they perceive as being the root of the delivery issues.

I worked with a particular CIO that summed it up well by saying "a grassroots effort will not be successful on its own. It has to have executive buy-in and support". I wholeheartedly agree with that statement. Unfortunately, this statement was uttered almost two years *after* the President of the organization had kicked off an Agile transformation in a top-down-mandated fashion by telling a combined group of over two thousand product and technology people that they were going to "be Agile". He had failed to do anything to change the environment around those people to allow them to achieve the intended outcomes. For the teams mandated to implement Agile it became all about checking the boxes. They were using new terminology and continuing to focus on output while not changing underlying behaviors to create an environment for success. They were *doing* Agile but not *being* Agile. By the time this CIO made that statement, the President had already been removed from the company, leaving a fractured workforce behind that had not yet realized any of the benefits that the Agile transformation was supposed to provide. By then the support and enthusiasm for the transformation had waned for many people, and there was an outright distaste for Agile by many others who came to see it as a process-heavy mandate that prevented them from focusing on their jobs and delivering more quickly. The project-driven, command and control nature of the broader organization had not been touched while the product and software development teams were given some basic training then thrown back into the fray of their current environment and culture. The executives were following a recipe for disaster from the beginning and saddled the rest of the employees with the less than desirable results that remained long after they left the building.

To unlock the power and meaning of Agile, executives and senior leadership must take on the challenge of focusing on *making* their people successful as opposed to *telling* them what will make the company successful.

> *"A leader is best*
> *When people barely know he exists.*
> *Of a good leader, who talks little,*
> *When his work is done, his aim fulfilled,*
> *They will say, "We did this ourselves"*
> Lao Tzu
> The Tao Te Ching

So as an executive or senior leader, how can you create that environment? How do you achieve the "culture shift" that so many others have aspired to and failed miserably? It is a matter of understanding the true meaning of "servant leadership". A true leader does not pursue a goal for glory and accolades alone. Through guidance, mentorship and creating an environment where people can stop merely surviving and start thriving, the primary aim and purpose will be fulfilled. Financial success and prestige will come of themselves in abundance when they aren't the primary driver of what you do every day. Becoming a servant leader starts with looking within and understanding the way you think about the company as a whole.

Holistic Agility demands executives and senior leaders take a "systems thinking" approach to the transformation effort. The organization itself is a complex and dynamic system made up of a unique combination of people, processes and technologies. Executives must have an appreciation and understanding of the entire system before attempting to improve it. By having a working understanding of each of the roles and feedback loops defined thus far, executives can then work to ensure that there is the least possible amount of friction between the components of the system. They can also ensure that those components, and the people within them, are well nourished through leveraging functions such as Finance and Human Resources to create the optimal environment that minimizes overhead and churn. These functions become the "shore" that shapes and supports the "lake".

FINANCE

P
E
O
P
L
E

L
E
A
R
N
I
N
G

MARKETING

While systems thinking provides the understanding of the current environment, executives and senior leaders must pair systems thinking with critical thinking to create a path to the future. Critical thinking asks, "how do you know what you know"? It leads you to question your long-held beliefs, including the underlying assumptions of your thinking. For example, why do you fund projects instead of teams? For many companies those types of basic assumptions are considered given and never questioned. By questioning your own thinking in a critical fashion, you unlock opportunities to achieve much more rapid and wide-spread transformation while using systems thinking to identify where to transform by finding root causes of delays between feedback loops and learning opportunities. Applying the two types of thinking together allows executives to improve overall company performance by optimizing the whole system of organizational behaviors and interactions in new and potentially innovative ways.

After setting the business strategy, goals and objectives and communicating them to the organization, one of the most challenging things for executives and senior leaders to do at that point is to get out of the way and let the people in the company do what they do. It is a matter of trust. As an executive, by trusting your people you can create a true sense of empowerment and purpose that increases productivity. By trusting your feedback loops you can release the need to seek some (false) sense of certainty of how the future will unfold, allowing you to focus on the here and now where the most important and valuable decisions need to be made. The feedback loops themselves provide continuous communications and the necessary sanity checks. By trusting the system, you can free yourself to drive the creation of an environment that enables ever increasing levels of success for everyone involved. Through that level of trust, you can release any remaining "command and control" behaviors and become a true servant leader that instills a positive psychology throughout the organization. So set your strategy, allow the teams to do their part, and use the resulting feedback, data and analytics to continuously mold that strategy while concurrently using the other tools at your disposal to create a supportive and healthy environment for the teams.

Executives and senior leaders that keep their eye on creating an environment that grows and sustains high-performing teams produce an enhanced level of adaptability to changing business needs. Having high-performing teams puts the company on the positive side of "Adapt or Die". It is less of a productivity loss when a high-performing team changes context or domain than it is to break up a team and put those people on other teams, effectively pushing *all* of those teams back to the "forming" stage. So, if you determine that you need to redirect investment levels to enable accelerated output in a particular area, moving an existing high performing team to that area equals the redirection of investment. Being able to redirect an existing team to a modified, or new, work context allows the company to gracefully accommodate events like market changes, disruptive technology changes, and new innovations. In short, you give yourself the opportunity to stay ahead of the game. Creating the appropriate financial and team structures to enable this organizational adaptability is where Holistic Agility requires coordination with Finance and Human Resources teams that will shape the environment in which these teams exist.

Finance

Focusing on project costs causes people to lose sight of what is valuable. It is not true that you can effectively determine where to invest funding based on an estimate of the cost to fully develop a feature, function or solution. Ziv's Law, also known as the "Uncertainty Principle in Software Engineering", states that "software development is unpredictable and that the documented artifacts such as specifications and requirements will never be fully understood". Any software estimation exercise that results in a predicted cost of development must include a guess at how many hours it will take a certain number of people to deliver given requirements. That estimated total number of hours is multiplied by some internal charge rates provided by Finance to come up with the estimated project cost. If part of the estimation exercise contains the result of multiplying hours and rates, and a long-lived high-performing team has a stable number of people (and therefore total hours available) at those same rates, then don't you already know *exactly* how much that team will cost for any given time period? It is time to simplify the finances and refocus the organization on delivering value early and often.

Almost needless to say, in a project-driven organization the projects are considered the basic unit of work. Documents are completed to define the project intent and objective with related scope and requirements. Estimates for project cost and duration are based on that initially defined scope and funding approval is based on those estimates. Finally, "resources" are assigned to the projects so work can officially begin. During the project, measurement is made against what were three inherently uncertain variables of scope, time and cost. It is textbook project management practice, but this all results in many costly impacts that are *not* included in the project estimates.

Using these three uncertain variables as constraints, commonly known as the "Iron Triangle" of project management, unnecessarily high levels of overhead exist for tracking what is actually occurring against all three of those variables, and when things change, more overhead is added when formally tracking those changes. The focus becomes fixated on protecting the plan and not adapting to change. The result is that the entire organization is either working to get the variables back in line against the original plan, or "re-forecasting" and "re-baselining" the plan when the actual results will not be contained by the original constraints. The entire company has forgotten that the first values for the three variables were all *guesses*.

This then leads to a divisive environment where people are pitted against each other to either get back in line with the original plan or be blamed for not meeting the original expectations. It must absolutely be called out here that what was a set of guesses has become documented directives and expectations. People get angry and defensive when expectations are not being met and there is a culture of fear created. Nobody wants to be the one blamed for why a project has gone off track, yet fingers get pointed all the time when reality does not end up matching the plan. Therefore, the next time an estimate is needed from somebody that was blamed for not meeting expectations in prior projects, they pad their estimates and give them plenty of contingency and padding to avoid future blame. That contingency is consistently used up in projects, but frequently on lower value activities because there is a built-in understanding that extra buffering was added to the estimates. The whole thing has become a game, but what that truly represents is lower throughput and productivity overall. It is costing the company *more* time and money to get the same things completed, and yet projects *still* slip into "Yellow" or "Red" status because no matter how much padding is added at the beginning, the unexpected occurs and new learning and understanding is uncovered during the project.

Red. Yellow. Green. When used in a project status report those colors only tell you if you are spending your funding at the planned rate. Only if the date is at risk or the project is costing more than the budgeted amount does a status report turn to Yellow or Red. "Project X is burning hot". "Project Y might go over budget due to the extended timeline". "Project Z is not going to be able to use all of its funding, let's move the extra dollars over here". Project financials are a repeating cycle of "burn hot, adjust to cool off, repeat". Schedule delays and budget overruns cause variances to be analyzed, resource allocations to be reassessed and projects to be re-planned. Impacts on other projects are determined, reprioritization happens, and resource planning is re-shuffled. Obtaining change control approvals can create even more delay. For anyone that has been involved in managing budgets and actuals across multiple concurrent initiatives, they know how exhausting all of this overhead can be. When looking at the churn these activities cause across projects and teams when trying to take corrective action, it is contributing to even more productivity loss. It also lowers morale because teams aren't staying together, and the team members are hesitant to voice any more concerns for fear of creating even *more* churn. They end up just coasting along while feeling disempowered by the environment they are in.

All of the efforts and impacts of protecting an inherently flawed plan has associated costs to the company that do not contribute to delivering the final product. If each one of these things were monetized when first assessing viability of a project, it would be clear that accepting all of those costs and impacts creates a less viable business case and a lower return on investment. All the churn and stress combine to create a less-than-healthy organization. Just as if you were being repeatedly thrust between a burning hot sauna and an icy lake, trying to survive the continuous and drastic changes to the environment when funding gets shifted around allows organizational and cultural sickness to creep in and take hold much more easily. To allow health and strength to build, the company can create a stable environment through how it approaches funding development efforts.

Moving away from a project funding model to one that funds long-lived teams immediately eliminates budget variances and surprises, removes the unnecessary overhead of tracking individual project statuses, and addresses the throughput and productivity issues. In the simplest sense, the shift is to **start to bring the work to the teams and stop bringing "resources" to the work**.

In the chapter on Business Strategy & Stakeholders we covered the behavior that spawns from having a project-driven annual budgeting cycle (which actually goes *all year long*), but what most companies shy away from is creating a funding model that funds capacity because they truly believe that it requires a change in their funding process. Very few people, executives included, seem to be willing to go there, but it is not actually necessary. It can be done with a change in thinking about how you ask for funding today.

Do not change your funding process. Change what you are asking for when you use that process. If you are pursuing an Agile transformation, and you understand the benefits of keeping teams together for an extended period of time and managing scope through a constant flow through roadmaps and backlogs, then it is now time to put your money where your mouth is. Fund the teams and stop funding projects. If the majority of the cost of a software development initiative is the cost of labor, and the number of people on staff is a known quantity, then why is there a need to continue to treat cost as an "unknown" that needs to be "estimated"? If you know the number of people you have, then you know what they cost. When seeking funding approval, request the cost of the team for a period of time, stating that the scope of work will cover any production builds and releases during that time frame. As was just shown, funding projects creates unnecessary overhead which limits the capability to change direction quickly and gracefully. Project funding inhibits basic agility while team funding enables Holistic Agility. Consider the following sample calculations:

- Each team member has a fully loaded hourly cost of $150 per hour
- Assuming 40 hours per week and 50 weeks per year, each team member has an annual cost of $300,000
- A team of 10 people costs $3M per year

No matter what they work on, that team costs $3M per year. That's it. People have worked in an unnecessarily complicated procedural environment for so long that they have become convinced the financial aspect of software delivery is more complicated than it needs to be. Why would an organization spend time deciding how to divvy up a $90M budget across a set of projects when that number tells you that you can fund 30 teams? Dividing a budget up into projects is the creation of something that does not need to exist. The actual cost to the company is not the projects, it is the cost of the *people* working on those projects. What gets highlighted as unnecessary and extraneous is breaking the budget up into a multitude of different activities such as projects, projects tasks, development time, testing time, etc., and tracking costs at that level of granularity. It creates an exponential amount of work for the organization that does not directly contribute to new and improved products and services. In fact, it is often an inhibitor to driving to the intended business objectives in the shortest possible time frame because of the process overhead required to "move money around". The people are here, and they define the capacity of the organization to deliver. Understand that you are paying for capacity and not projects, and let the teams do what they do while learning and adapting constantly.

Another way to think about the shift to funding teams is that the intent is to move from having to manage the three project management variables to having to manage only one. By realizing that the cost of a team is known, and the team will develop on a constant cadence, then the variables of cost and time are effectively eliminated from the equation. The only thing left to manage is what the teams are working on at any given time, being the scope, which needs to be something that can change at any moment through the reordering of the backlog. Whereas in the project model there is a low level of consistency and certainty, in the team model the level of cost certainty *increases* because the burn rate of the team is constant. Achieving cost certainty through estimation will never happen.

For companies that place a value on and depreciate software assets, there is one aspect of funding that needs to be considered and addressed for legal, audit and financial reporting purposes. For people that have worked in these types of companies, they will most likely know it as the need to differentiate Capital Expenditures (CapEx) from Operating Expenses (OpEx), or simply capital versus expense.

From a legal standpoint, the law is fairly simple and straightforward. Software built for internal use, meaning software not marketed and sold to customers outside of the company, can be recognized as a fixed asset that can be depreciated over time. The benefit is that depreciation lessens the tax burden of the company.

To identify what costs can be included in CapEx versus OpEx, there are 3 stages of development:

1. Preliminary
2. Application Development
3. Post-Implementation

Preliminary costs are all charged as OpEx and include the time spent making decisions on resource allocations, determining requirements, and evaluating technologies and suppliers. Application Development has both CapEx and OpEx categories, with any data conversion, training or administrative costs being OpEx while all coding, testing and hardware installation costs can be capitalized. To calculate the amount to be capitalized the company can include the payroll costs of the people included in the development effort as well as any materials and services used in the development effort such as software purchases and third-party development fees. Post-implementation costs are all expense and include training and maintenance expenditures.

This maps very well to the team funding model:

1. Preliminary – team creation and funding, roadmap and backlog management including investigative and exploratory efforts such as "spikes"
2. Application Development - time spent completing and delivering new features, functions, enhancements, stories, etc.
3. Post-Implementation – maintenance and production support, including production defect fixes

Therefore, a team member's timesheet used to track actual costs could be very simple to fill out by having one bucket for new development work, one for expense to record backlog refinement, spike and defect work, and administrative buckets for paid time off, company meetings, training and the like.

From a budgeting perspective, the method of funding teams will be determined on if there is a single overall spend budget, or separate CapEx and OpEx budgets. For a single budget, the total amount can be used to fund a bucket for expected hardware and software costs (capitalizable) and use the remainder to fund the teams. If the company's budgeting process starts with different summary budgets for CapEx and OpEx, you can use the total amount of CapEx and OpEx provided to each portfolio, program, product or functional area and convert that total amount into a headcount number, again initially setting aside funding for expected hardware or software costs first. This remaining amount will tell you how many teams you can fund out of that total combined budget.

For example, a department responsible for the development of the company's point of sale solutions receives their allocation of the overall budgets as $9M for CapEx and $4M for OpEx. If each team costs $3M then there is enough total budget to fund 4 teams in that area with $1M available to cover hardware and software costs.

To manage the work under those budgets, the prioritization and sequencing of work in the backlogs becomes a powerful tool. If your teams have done extensive work on new features and have been spending CapEx at a faster rate than OpEx as a result, then the decision can be made to move more maintenance work and defect fixing to the top of the backlogs or pipelines to recover on the OpEx side or vice versa. That becomes an interesting scenario to discuss with your Finance group because, given the fact that CapEx reduces the tax burden of the company, going "over budget" on CapEx while "underspending" on OpEx might be a positive outcome. I worked with an organization that had not yet fully moved to a team funding model, and at the time were touting the amount by which they had unexpectedly exceeded their CapEx spending target, to the tax benefit of the business, but in reality had left many more capitalizable dollars on the table because they were only capitalizing portions of CapEx defined and funded projects but not all of the new feature development work they were performing in addition to those projects.

No matter how you get there, funding teams instead of projects minimizes staffing volatility so that team productivity and performance can improve, reduces the cost of overhead, and it will eliminate most of the unfortunate behaviors mentioned in the chapter on Business Strategy & Stakeholders. There is an added benefit of funding teams in that your overall budgeting and funding process to become Agile as well.

Instead of a project-driven budgeting and funding model, a team capacity funding model creates the capability for adaptive planning and budgeting. When dynamic market changes or new learnings dictate that investment levels should be adjusted to quickly exploit opportunities or reduce costs, the executives and senior leaders have the capability to adjust capacity by reallocating *teams* to areas where the use of that capacity is better suited. They do not impact team productivity dynamics by moving individuals, which would send multiple groups back into the forming and storming stages. It becomes critical, yet unique to each company, that they have a clear definition of what constitutes a self-sufficient capacity unit that can be moved when necessary. It is based on the cross-functionality discussion in the chapter on Development Teams, ensuring a given team includes all the skills necessary to remove outside dependencies and time-box their development efforts. When the unit of funding is a team, then the company can determine that capacity can be reallocated to an area that is thriving while another is diminishing or identify incremental funding to build new teams to expand capacity for areas in need. Having a capacity-oriented approach to financing and budgeting can be one of the biggest enablers of Holistic Agility, allowing teams and the company to course correct frequently with minimal friction, and is a significant contributor to creating an overall environment oriented to business outcomes versus team output.

Human Resources & People Operations

"HR is getting involved". That is an ominous statement for almost anyone that has ever heard it. Human Resource departments have traditionally been focused on making sure the company and its employees are compliant with the law, and work to reduce the liability of the organization, especially the avoidance of litigation. This may have made business sense when both companies and employees expected to have long-term relationships, but today it seems almost laughable to anyone that they would work for the same company their entire career. The power has shifted from the companies to the employees. They control their own destiny and will readily leave a company if it does not provide them with an environment in which they can grow and thrive. With one of the main objectives of Holistic Agility being to keep teams together for an extended period of time to achieve the "norming" and "performing" stages of the Tuckman model, it becomes imperative for the organization to do what it can to obtain and retain top talent.

Human Resource departmental titles are showing the drive to a new employee-centric focus by changing to things like "Employee Experience" and "Talent Management". Google no longer uses the term "Human Resources" either. They now have "People Operations" and the motto of that department is "Find Them, Grow Them, Keep Them". For purposes of this discussion, we will stick to using the People Operations term.

I have an Agile Transformation consulting colleague, Elisabeth White, who champions people at every level in every client, even internally to our consulting company itself. Any time someone uses the term "resource" when talking about people, she calls them out. "They are *people*, not *resources*". It is an important distinction when determining how a company views and treats its employees. When you don't see them as people, thereby not recognizing that they are all equal yet unique human beings, it is easy for the entire organization to slip into an "order taker" mentality and not realize the human value and potential of every single person in the company.

Regardless of what title is on the department, they have an enormous opportunity to influence the team dynamics throughout the organization. Through their hiring efforts they contribute to deciding who is on the teams, and through establishing programs for things such as training, mentoring, career paths and recognition they contribute to retention.

People operations can work with leadership across departments to ensure each team's purpose is clear and that funding and skills set requirements are appropriately aligned. If the number of people with requisite skills do not currently exist on a given team, then people operations can assist in filling those gaps by understanding where those skills can be found or grown in the organization. While it is easier with training and coaching to grow skills of existing staff, and less costly, the other option is to perform the recruiting and hiring functions as they do normally. With team dynamics being paramount, People Operations does not necessarily make the final call on who makes the cut.

A department dedicated to the people in the organization is responsible for creating and maintaining teams that are happy family units. One of the behaviors that happens at many companies is that a job description is written up by a manager and the recruiting and hiring is driven by HR and a few select interviewers. When driving for team cohesiveness, this type of hiring process begs the question, why aren't the TEAMS doing the interviewing and making the hiring decisions together as a unit? If the organization is willing to take the productivity and performance hit on a team by altering its membership, then the entire team should be identifying who they think will be a good fit for their already tight-knit family. They are in effect choosing a new friend and family member, so they should have final say on who gets added to the team. That applies to people joining the team from other parts of the organization as well as new hires.

One of the balancing acts of the people operations group is how to manage individuals that are damaging to the team. For productivity's sake, as well to minimize the collateral damage of psychological contagion, removing toxic team members without delay or remorse needs to become the norm. When a foreign substance enters the body, it gets immediately attacked and purged. Your body doesn't wait around to see just how sick you can get first. Creating and preserving a positive team dynamic minimizes the damage and productivity loss to the organization.

Another aspect of creating the right team dynamic is to have appropriate compensation models in place. Compensation models that, when hiring top talent, the company can pay them enough to take the money issue off the table for the individual, allowing them to focus on doing quality work. This is a critical concept when it comes to enabling top performance. Once someone is no longer concerned about meeting a minimum pay threshold, what motivates them is not more pay or large bonuses. People are actually motivated by having a clear purpose, being able to work in an autonomous fashion, and to pursue increased mastery of their chosen skills (more to come on this in the next chapter). When the individuals are motivated appropriately, the people operations department can focus much more on establishing and preserving the team dynamics over risk and liability avoidance. The term "HR is getting involved" becomes a positive because their objective is to make the people happy enough to be intrinsically motivated which contributes to increased productivity.

Part of the responsibilities for people operations can even be working with facilities management regarding the physical workspace in which people spend their days. One of the outcomes being pursued is to facilitate social connections between employees. By building co-located teams and placing them in an area that has both collaborative spaces as well as places to focus on individual work (such as a "caves and commons" office space setup) you are enabling human interactions to occur more naturally and team bonding to happen continuously. Beyond the physical environment of each individual team, people operations can facilitate introductions of people across the organization through various programs, events and activities which creates a sense of community and belonging that does not happen when people work in isolated environments day after day. The goal is to support every individual and make them feel connected to something bigger, be it their team, their department, or the company overall. By keeping a continuous eye on how to best support individual needs, the employees naturally fill in the gaps to make the overall organization continue to operate at its best.

A great example of creating a connected family environment, albeit not the most positive circumstance, was something I was able to participate in when working for a company many years ago. A particular employee had a tragedy happen in his life and had to take some extended time off to deal with it appropriately and create a completely new support structure for his life outside of work. The HR team at the company created a policy that allowed employees to donate some of their own vacation time to him so that he could take the time necessary and without worry. From the company's perspective, the vacation time on the books remained the same but it made a huge difference in that person's life, and for the rest of us we felt a new connection to each other and the company for making that possible. Focusing on what is right for your people, as opposed to treating them as "resources", can generate a collective commitment to each other and to the business.

Another area where people operations can have significant impact on social connections is through recognition and incentive programs. Historically incentives have been at the individual level with an "Employee Of The Year" feel, or "Coming Through In the Clutch". But frequently those types of incentives do not recognize the collective contributions made to achieve business success and may even result in divisive and subversive behaviors when others feel slighted. By changing programs to recognize and incent teams instead of individuals, it supports the type of team ownership of outcomes that has been highlighted in previous chapters. No longer do you want to reward and perpetuate any "hero" behaviors when the fact that a hero had to step up oftentimes points to an improvement opportunity. It is also important to identify if any incentives are date driven, as we are moving away from a project plan-driven environment to an adaptive and Agile one.

Incentives for teams can be at the individual team level or common to the entire organization. Every single team member needs to be incented around driving business outcomes, such as KPIs. Output-oriented incentives, such as increased velocity or number of features, or being "on time" and "on budget" for projects, are disincentives from an organizational perspective because it takes the focus away from doing the right thing right now to drive the business outcomes. By tying incentives such as bonuses to KPI achievement, or even doing something like a progressive pay-out scale on increasing percentages over the target KPIs, could be a shared reward system because the working assumption is that those KPIs are all rooted in achieving financial goals. The other thing to keep in mind is that the incentive and reward systems could change frequently because new KPIs may become more relevant or the external environment of the customers and market may change. People operations needs to be as flexible, adaptive and agile as the rest of the organization and have the freedom to support people continuously during constantly changing circumstances.

As a side note, while a company is on a transformation journey the wait time for learning will be diminishing. Since the company *wants* that wait time to diminish, People Operations has another potential set of incentives that can be put in place that are based on continual improvement initiatives. For example, a company could decide if completing improvements that allowed what was previously only quarterly releases to happen monthly warrants some type of reward or recognition because the delay in learning to the company was just reduced by 60 days. From an executive level perspective, being able to drive to business outcomes while optimizing the entire system are both areas that warrant incentives as the two together improve overall performance. That said, you can also get creative with rewards and recognition depending on the current state of the organization. You could do something like an immediate recognition that is triggered at learning points, such as recognizing that "the release that Team X just launched had immediate positive impact on our top KPI". The possibilities are endless and the more people that share in the success of meeting the business objectives, the less grumbling and blame occurs between individuals, teams and departments. If teams' incentives are tied to the same KPIs that the entire organization is driving towards, people operations has created an alignment in purpose, and an opportunity to collectively celebrate success, that permeates the entire company.

Organizational Learning & Maturity

Muscle Memory. It is a term discussed earlier when working to get new behaviors ingrained in development teams so that, when under stressful circumstances, there is a graceful and predictable reaction to a changing and chaotic environment. Regressing back to old behaviors is a constant struggle with Agile transformations. It is a natural human behavior to revert to deeply ingrained responses when under stress. Neurological studies have shown that stress hormones inhibit activity in the decision-making and goal-oriented areas of the brain, but not in the areas involved in habitual behavior. Therefore, it is easy for individual teams to slip back into old habits if Agile processes have not been established as "muscle memory" for the entire organization. That means the time and effort spent trying to transform the organization is at risk of being wasted. That means there is an inherent need to institutionalize organizational learning for both sustainability of the transformation and continued growth in the future.

"Ancora Imparo"
Michelangelo

The quote "Ancora Imparo" has been attributed to Michelangelo when he was 87 years old. It is an Italian phrase meaning "Yet, I am learning". After achieving unparalleled success through a creative career that changed the world, Michelangelo knew that there was no end to learning new things and making yourself better and better every day. For Holistic Agility to take hold in an organization, there must be mechanisms put in place that support a culture of learning and continuous improvement.

In a similar fashion to the efforts of people operations in creating a social support structure for employees, one of the key support mechanisms in an Agile environment is the establishment and perpetuation of Communities of Practice (CoP). A CoP is a communication and support mechanism for people with common interests across the organization. Early in an Agile transformation, these groups may simply need each other for moral support while the environment is changing rapidly. People operations can help communicate when new CoPs are being initiated to help drive participation and attendance in an effort to support their employees through the change. As the organization progresses, it becomes much more about knowledge sharing and collaboration in an effort to improve everyone's effectiveness and experience, so CoPs may start to spin up in a much more organic fashion. The specific CoPs that need to be initiated will be defined by the company's current state and what pain points exist at the time. CoPs can be:

- Role based, such as Scrum Masters and Product Owners
- Skill based, such as iOS coding or artificial intelligence (AI)
- Topic based, such as DevOps or Acceptance Test Driven Development (ATDD)

For example, if there is an initiative to move from manual testing efforts to automated testing, then a Test Automation CoP may be in order. When standing up a CoP, identify champions and progressive thinkers in the organization that have an interest in that area of expertise and invite them to be the initial chairperson. They can help drive awareness and attendance, knowing that their passion on the topic is often contagious, and they may already be looked to by others based on their knowledge and experience. As time goes on and attendance grows, sharing responsibilities and rotating roles in the CoP will promote the sustainment of the community until there is no longer a need for that CoP in the organization.

Underlying this evolutionary approach is an assumption that the CoP is completely self-organizing. The membership is not mandatory, and the agenda is defined by attendees' collaboration on what are their current pain points and hot topics. Any type of management control or dictation of content in a CoP will not provide the right environment to achieve the intended goals. Everyone that walks into a CoP meeting is on equal footing and is free to raise and discuss topics and ideas with no fear of judgement or retribution. An environment of trust will allow the true issues to rise to the surface quickly. Communities of Practice, in addition to team retrospectives, can be some of the most fertile grounds for identifying improvement opportunities. When a CoP is self-organizing it creates a psychologically safe and supportive environment where individuals can raise any type of issues and brainstorm ideas on how to make things better. This is the point where a CoP needs its own support structure as there needs to be another mechanism to turn those ideas and opportunities into reality.

Separate from CoPs, the primary task of a Center of Excellence (CoE) is to drive and implement change. This is another group that may go by many names in different organizations, with any combination of terms such as Agile Transformation Team, Steering Group or Agile Learning,

Coaching and Training. A CoE is responsible for broadcasting, implementing, and sustaining transformational successes identified through individual team experimentation. They drive continuous improvement and learning through experimentation and knowledge sharing. The CoE is the group that is engaged with various teams, departments and Communities of Practice to understand where the current pain points and opportunities are for improvements, which can then be prioritized, piloted and subsequently rolled out to the entire company when proven successful.

The Center of Excellence can also be another powerful tool for use by the executive and senior leadership teams. In fact, it is best when the CoE is directly accountable to the senior leadership team. The function loses its power and value when embedded in a particular group, such as Technology or Product, because in those situations the group does not have the inherent or implied authority that elicits active participation from the entire organization. Shortcomings with the current environment, such as too much time between dev complete and production release, can be put on the CoE roadmap and backlog to be addressed and the senior leadership team can function as the stakeholder group that prioritizes CoE work to ensure the people on the team are always working on the improvement initiatives that are the most valuable to the organization in terms of achieving their next level of maturity, efficiency and effectiveness relative to the current business objectives. The CoE can then work with various teams, CoPs and departments to coordinate improvements, perform isolated experiments and promote successes to the broader organization to institutionalize them, effectively working to put them into "organizational muscle memory". It is a never-ending cycle creating a culture of life-long learning and relentless improvement.

One of the key factors that helps to ensure the success of the CoE is that the presence of its members must not be forced on the teams or CoPs. The CoE must be viewed as a group of trusted advisors that coach, guide and improve the organization, but more than that, they listen effectively and without judgement. If the CoE is not seen as the trusted advisors that can help make the environment better for the teams, then transparency and honest communications will be diminished. Many organizations have taken the approach of allowing the teams or CoPs to invite CoE members to participate so they may engage on their own terms. Regardless of how the CoE relationships are approached, the goal must be to bring various people together to share learnings and extend their influence across the wider organization.

Institutionalizing organizational learning and maturity creates the connective tissue needed between the various groups in the organization. That connective tissue supports, connects and binds the other departments together and allows them to get stronger and stronger over time. It is also lightweight in that the CoPs are self-organizing and self-sustaining, and the CoE can support even a large organization with relatively few people. It becomes the equivalent of white blood cells in the body that identify sicknesses within the organizational organism. Without them, toxicity begins to build and cause damage to everything around them. Allowed to do their job effectively, they deal with issues accordingly to preserve and improve overall health and well-being.

Marketing & Sales

"That guy could sell ice to an Eskimo". For many companies their marketing and sales initiatives are driven solely by the profit motive. This turns much of their methods and communications into various types of fear driven attempts to make consumers buy something they don't necessarily need. The traditional focus of these groups is on the competition, the money, the top down sales quotas, the quarterly numbers, and a multitude of dollar-oriented objectives. Using those objectives as the main organizational motivators can lead the company down a dark path.

There is a publicly available video from Dan Pink that summarizes his book titled "Drive: The Surprising Truth About What Motivates Us". In that video he discusses that money, often combined with a fear of punishment, is not the best motivator for improved performance when completing jobs that require thinking skills and creativity. In those cases where decision making is required, higher pay counter-intuitively creates *lower* performance. What really motivates people is the desire to be self-directed, to improve their skills and to do something that is meaningful and important. Pink summarizes these three as autonomy, mastery and purpose. People are moved to higher levels of performance and satisfaction when working for companies that have a transcendent purpose. He rightly points out that when the profit motive has become unmoored from the transcendent purpose that "bad things happen".

The correlation here is that, when marketing and sales teams are profit and fear driven and not purpose driven, they not only perform less effectively, they are often incented to do things that put the company and its customers at risk. Similar to setting up self-organizing development teams that are constantly learning new skills and delivering value to the customers, that same application of purpose, autonomy and mastery can apply to the teams that are seeking out new ways to reach new customers.

The chapter on Business Strategy & Stakeholders discussed how to incorporate and prioritize inputs from the entire organization and to engage those stakeholders in reviewing iterations of the product as it is being built. Marketing and sales departments should be directly involved in those activities as well because they are excellent sources of information that can drive prioritization of future product functionality. They know what the market trends are and what sells. I worked with an organization on a transformation where we brought the various sales and business development teams together to meet with the product organization in a recurring fashion to discuss what was causing problems for sales as well as ideas of what new functionality they believed would allow significant revenue increases to be achieved. Product incorporated that feedback into the overall discussions with the broader stakeholder groups when reviewing what had been built in the last iteration, and then decided if the roadmaps and backlogs needed to be re-prioritized for the upcoming development work. Until that point the sales and business development teams were not engaged in the process unless they sold something that didn't exist in the product and the development teams had to scramble to meet the obligation that they had unknowingly been committed to. Once the teams were regularly engaged the frequency of those types of "build if sold" fire drills started to drastically decrease. It was simply a matter of communicating and considering every voice as an equal stakeholder when sitting around the table to prioritize the work of the development teams.

By incorporating marketing and sales people into the development life cycle, it also provides the opportunity to align their goals with the KPIs and scorecard items of the company. That flips the scenario from leading with the profit motive to leading with the transcendent purpose. The thought being that if every employee, marketing and sales included, is aligned to and living the company's purpose then abundance will flow. This is even applicable for companies that choose the OMTM (one metric that matters) approach, as sales targets to be measured against will be different depending on the type of metric that matters to the organization at that point in time. Continuously driving goals simply by things like dollar quotas while the rest of the organization is attempting to provide customers with value and learn quickly results in undesirable behavior. It does not motivate the marketing and sales team to truly identify with the customers and help them meet their needs. The goal from a Holistic Agility perspective is to create a mutually beneficial relationship between the company and the customers.

When marketing and sales teams are aligned to the company's transcendent purpose, they will continuously be asking themselves "what is it that our customers need?" There will no longer be a drive to sell customers things they do not want. There will be an increase in "truth in marketing" as opposed to "fake news" types of campaigns. Finally, marketing campaigns will begin to function in an Agile fashion themselves, as the duration needs to be shortened so that the company can learn quickly that what they *believe* customers want is truly what they *do* want.

There is another interesting potential outcome when companies remain holistically focused on their transcendent purpose and not on finances. It creates the possibility that more companies will not go the route of becoming publicly held because being publicly held drives behavior that is solely driven by achieving quarterly growth, often at all costs. There becomes the potential, which is already trending in several industries, for more companies to remain private or convert to employee ownership. An Employee Stock Ownership Plan (ESOP) is a viable business structure that speaks to the goals of continued independence, employee retention and allows the focus and the organizational culture to remain laser focused on the symbiotic relationship between the company and the customers. This would be a further enabler to allow traditional marketing and sales teams to align with the intent of Holistic Agility.

What Holistic Agility Is Really All About

Learning is a process of incorporating change as a result of experience. Holistic Agility is about continuously learning and growing. It is about experimenting with new ideas and quickly identifying what works and what does not. It is the scientific process in action while you are repeatedly testing out different hypotheses. To do this effectively the whole organization must be working in harmony. So, let's return for a moment to the discussion on why you want to be Agile.

It is when you learn that what you're doing is not the right or most effective thing to be doing to achieve your goals that there is an immediate need to be Agile. The entire organization needs to be able to change direction gracefully. Looking back at the dictionary definition of Agile, it is "marked by ready ability to move with quick easy grace; having a quick resourceful and adaptable character". Agility is about learning and adapting, but speed has always been part and parcel of any discussion on Agile transformations. So much so that many people now directly equate the term "Agile" with developing software faster.

If you want to deliver "faster" you must make concerted efforts to eliminate waste, dependencies and wait time in all of your processes. To be faster you are striving to apply the concepts of Lean that have most often been applied in both manufacturing and startup contexts. Taking one more foray into the dictionary, the term "Lean" is defined as "characterized by economy (of operation)". From a Holistic Agility perspective, this means driving toward the shortest and most efficient path between coming up with a new idea and getting it into the hands of the customers. The problem is, being fast unto itself is not valuable either. The goal is not to "get it done *faster*", it's to get it done fast *and learn if it was the right thing to do.* The company will be learning if the solution provided to the customers has achieved the desired outcome. This is the combining of speed, agility and learning into one.

The mindset being established through Holistic Agility is one with a unified focus on desired outcomes. It is not concerned with the mere delivery of projects. All initiatives, projects, objectives, KPIs and metrics for any business are rooted in terms of revenue (generating and/or protecting) and cost (efficiency and/or reduction). The company's *products* are what generate revenue and incur costs. *Projects* are a way to create or modify those products. By establishing a clear organization-wide common understanding of what your products are you can then use that as the basis from which you build your teams. The products themselves will define all of the skills a fully cross-functional team requires to design, build, test, deploy, monitor and maintain those products. Once those skills are identified you can put people's names to those skills, remembering they are people and not resources, and that they will come together to create their own dynamic through which solutions will emerge. The longer you keep these people together, the more productive, cohesive and higher performing they will become. If you treat them as "resources" whose time can be allocated, leveled, and pivoted left and right, up and down in a spreadsheet, then you've lost sight of the power of their humanity as a collective team. Holistic Agility is about the continuously increasing speed, agility and learning possible through the collective power of the people in your organization.

That collective power can be harnessed to create a win-win-win scenario. When the team members believe in and support each other, and they know that what they do every day is valuable, then they are enthusiastic and happy with their jobs. That's an employee win. Those employees create valuable and needed solutions for the customers. That's a win for the customers. When those customers use, and refer others, to the solutions then the company gets the win. When the company shares those wins with the teams, it creates a virtuous cycle that is hard to break.

There is one word included in the last description of a team that needs to be highlighted. That word is "happy". Happiness is the predecessor to a successful, hardworking organization. Hard work does not create success that then leads to happiness. This is the misnomer of many an organization and has been proven to be the key to success for many of the most admired companies. Purpose, autonomy and mastery are what motivate people in the here-and-now. So-called "success" through hard work does not provide a sense of purpose, autonomy and mastery. It is the difference between treating your employees as "resources" used as a means to get work done versus treating them as the wonderfully unique individual human beings they are who then come together to achieve the magnificent. It is the difference between creating a culture of fear, where no one dares to speak openly to their "superiors" and creating a stable environment with solid support and social structures for the employees that enables a positive psychology that permeates the entire organization. Holistic Agility aspires to transform the company into one that is happy, healthy and whole.

The Playbook – Managing Change

Embarking on a meaningful transformation is an act of courage. The path taken by most is what many people would call the "easy way out", avoiding the difficulties of changing while succumbing to the current state and becoming jaded, cynical, stressed and apathetic. Calling that "easy" couldn't be further from the truth because the pain for those people persists on a daily basis while the clear reasons for change are rationalized or ignored.

The only thing that is constant is change. It cannot be avoided. It is less disruptive to embrace change than being averse to it, but, recalling the snake skin analogy, change is inherently uncomfortable. If people could become progressively more comfortable *being* uncomfortable, then change becomes easier because of the reduced resistance. Unfortunately, resistance to change is the norm and the organizational pains continue to linger and grow. Once the pain has become too great for someone with the drive and wherewithal to initiate meaningful change, an organizational awakening can begin.

If that change agent is at the executive or senior leadership level, they have the luxury of starting with some of the "big rocks" like creating co-located, long-lived stable teams and funding them as such. Once they have created an environment that is effectively nourishing the teams' success, they can turn their attention to improving company performance by identifying what keeps them from learning every day. A very effective technique in this context is value stream mapping, which clearly highlights where wait time exists in the business process flows and covers the entire "concept to cash" business processes. Unfortunately, true transformations (which do not include mandated team level changes) rarely come from the head of the organization, they come from the heart.

With the teams being the heart of the organization, setting the "heartbeat" cadence of delivery, they are where true transformation is more likely to begin to take shape. The "drop in the lake" corollary is predicated on the assumption that there is sufficient environmental need and management support (blame?) to begin working at the team level to initiate *some* form of change. Whether coming from the head or the heart, the transformation is never going to be an overnight success. It is evolutionary and in the early stages will be a test of patience and perseverance. You are always starting from the current state, the assessment of which is the first page of the Holistic Agility transformational playbook.

When assessing your current state, it requires blatant honesty without judgement. You are where you are. No blame. Any delusion here leads to merely managing symptoms while not identifying and eliminating systemic root causes. It is the path to "Agile In Name Only" which is the equivalent of a transformational failure. To truly understand the current state as your starting point, question why things are the way they are and talk to people across the organization to get their perspectives on how things came to be the way they are. Along the way you can also identify the pain points each person has today by asking them what would make their lives better or easier.

Once you have a deep and thorough understanding of your current state and how you got there, you can define an initial future state goal that will be achieved through transformational initiatives. By taking the entire set of pain points identified through the current state assessment, identifying the part of the organization where the root cause most likely is for each one, and then mapping them to the concentric circle feedback loop diagram you can identify where to start experimenting with small changes. Through this mapping exercise you will also have in effect created a backlog of transformation items that can be managed with a CoE just like a team backlog as you learn what changes are effective and what is not. The future state goal you define at this point will state exactly how the entire organization will be able to learn more quickly than they do today. That may be in the form of fewer defects found in production or being able to perform A/B testing to determine where to pursue further development. It could be anything that is relevant and valuable to the current state and purpose of the company, but it needs to be clearly defined so all transformation initiatives are aligned to that outcome. It becomes why you want to be Agile right now, but it must also be understood that this first "future state" objective will only be your first milestone on the journey. Momentum will need to be established to enable continued success and growth through multiple layers of transformation.

At this point communication and transparency will be vital to the success of the transformation. Once you have an initial goal defined you need to share it with the *entire organization*, even if it is just for socialization and awareness purposes initially. Being able to articulate how these changes will address current pain points makes the transformation efforts a value-based initiative with its own business justification. Knowing that you are driving to a much longer-term objective of Holistic Agility means you can use this initial set of communications as an opportunity to plant seeds in people's minds that will be cultivated over time as you ripple out through the rest of the organization. This is also an area where formal Organizational Change Management (OCM) approaches and expertise can be highly valuable by preparing, supporting and guiding people and teams through the transformational change.

The definition of your longer-term holistic transformational goal (which will be the reason why you eventually need the entire organization on board) must also be aligned and oriented to the mission of the company. By being able to articulate the company's mission as its reason for being, such as supporting X market with products and services 1, 2 and 3, you have in effect captured the equivalent of the company's transcendent purpose. A transcendent purpose looks beyond the profit motive and focuses on providing something that makes the world a better place for customers. The seeds of almost all entrepreneurial endeavors are rooted in filling an existing need in people's lives. No matter how long your company has been in existence or how large it has grown, it came from an attempt to solve a problem and meet a need.

Through the Holistic Agility transformation, the goal is to learn every day if the organization is fulfilling its purpose in the most efficient and effective manner. By working your way through each concentric circle of feedback loops, each department's role in supporting that purpose becomes aligned with those around them, eventually creating a collective drive toward a greater good. Aligning short term change initiatives to this long-term goal will keep the entire transformation effort focused and minimize attempting change efforts that may be well known in the industry, but do not necessarily apply to your company and its purpose.

Now it is time to put the organization to the test. Take what you have learned through the current and future state exercises and begin your first change initiative at the development team level by establishing a basic understanding and implementation of Agile as the first drop in the lake, identifying a lightweight method or framework that will work well in the current environment. Provide the team with someone that has the skill set and some practical experience using the Agile methods you select. If those skills and experience don't exist within the organization today, then it would be useful to hire someone onto the team that does, or add an Agile coach to the team, to provide in-the-moment feedback and guidance until the methods and behaviors start to become second nature for the team members. The initial goal is to establish, support and celebrate development team improvements. Sending team members to a training class then sending them back out into the wild to put it into practice does not work because they are primed to revert to old behaviors and habits as soon as the pressure of delivery is felt.

Although the first change initiatives are intended to benefit the capability of the teams to deliver, the role of product management must also be part of this initial step as the overall team's success depends on having a good flow of work going through the pipeline, roadmaps and backlogs. That said, one of the most valuable fundamental concepts to establish and communicate early on is that these initial changes are intended to create a "pull" system where the team will define how much work they will commit to delivering in a given time frame. It is the basis for creating an environment of empowerment and decentralized decision making. Anyone other than the team that defines how much work will get done in what time frame has placed the organization right back in a project-driven mindset without actually having a project defined. Establishing a pull system is critical for overall transformational success but is one of the most challenging things to do early in a transformation because not being able to tell the team exactly what to do and when to do it is one of the hardest "command and control" behaviors for traditional managers to give up. It would be wise to include in your communication plan that management will need to focus on how to give the team the environment and tools it needs to succeed while giving them the time and space needed to establish "muscle memory" and improve their pace, often while going through the forming and storming stages if this is a team that has been assembled recently.

Once you have your first team (or teams) showing quality benefits of the initial changes, you can start to refine their performance through items identified through retrospectives and your transformation backlog. When a team is in the early stages of their Agile journey, the immediate pain points are often focused on how to better prepare work items for the teams, so they are ready to be pulled from the pipeline when team capacity becomes available. Working with both product management and the development teams on pipeline flow issues, such as how to elaborate, refine and decompose roadmap items into useable backlog items, will highlight that you are prepared to ripple out to the broader organization. As you do ripple out through the various feedback loops, groups and touch points, understand *why* the people in the adjacent group(s) would care about changing, or what their pain points are that would give them reason to change. It will help facilitate the initial conversations more effectively and help you to determine what type of tools, techniques and concepts to contextually apply.

From a conceptual perspective, the transformation initiatives will be applying a variety of methods, include some Agile concepts as well as some Lean concepts. Progressively identify where the friction and pain points move to after previous changes are in place and utilize whatever methods and tools will remove that friction and highlight the next friction point or bottleneck. It is a progressive improvement initiative that, once a change is made in one area another one will be highlighted as the next step to take to smooth out the touchpoint between the two. So, as product and development teams improve their work intake, prioritization and feedback loops, the need for change at the business strategy and enterprise architecture will be likely to be highlighted as the next optimization need.

Progressing a transformation to the business strategy and enterprise architecture levels becomes a slower and more deliberate transformation process that needs a much more systemic set of issues to be clearly articulated to the leaders of those teams as well as executives and senior leaders. When the ripples of transformation reach those outer circles and begin to slow momentum, more force must come from behind to overcome the inertia. By continuing to do more and more transformational improvement initiatives at the team and product levels, you are in effect dropping more rocks in the lake and forcing more pressure toward the outer reaches of the organization. The force being created through each of these subsequent ripples will provide you with clear reasons why the broader organization needs to change to allow further growth and maturity to occur for the benefit of everyone involved. You will be glad the seeds were planted through your original communication plan, and through the broadcasting and celebrating of other successes along the way, because now will be the time to harvest what those seeds have produced. The outer circles of Holistic Agility transformation are much more about Lean concepts, allowing the company to build, measure and learn as quickly as possible. With the active participation of executives, leaders, strategy and architecture, the duration of that cycle can be continuously shortened. Ultimately the point the transformation is driving toward is to:

- Understand customer behaviors and needs
- Determine how the company can best support the customers
- Provide solutions to customers in the shortest time possible

In other words, everything the company does should reference the customers in some form or fashion.

The Customer Journey

Many of the transformational improvement initiatives discussed so far have been directed at how to make the company more effective and efficient at providing solutions to customers and being able to observe if those solutions are having the desired business results. There is a risk of putting too much focus on optimizing the flow of work to the teams and not balancing that with a customer-centric view of the actual work to be done. Individual features, functions and stories can become less valuable to users if they lose their context to the overall customer experience that exists when customers utilize the product to accomplish whatever it is they are trying to do. To mitigate that risk, and to provide organizational focus on customer satisfaction, an activity that can be added during a transformation is to visually represent how the customers use your products. Capturing the customer journey and using it to determine what work to perform next can bring focus to every group in the organization that contributes to delivery. It can smooth out the remaining rough touchpoints between teams and make the Holistic Agility lake placid.

One of the most effective tools development teams use to capture the customer journey is user story mapping, while others use journey mapping when approaching things from a customer experience (CX) lens. User story mapping is a technique that provides context to backlogs. By first visually capturing a high-level sequence of events that users progress through when using a product, the features of the product become represented in the form of a user flow. User stories, captured as various functionalities within each given feature, describe the details and options that a customer may have when at that stage of their journey through the product. This method can be used to define a minimum viable product (MVP) of a new solution or capture the currently available functionality of a product as a user story map. The same type of content can be captured in a Customer Journey Map, which will also layer in customer thoughts, feelings and reactions. In either case, ideas for new features and functionality can be inserted into the maps to define what changes to the product would make it more valuable and easier to use by customers. It allows you to easily identify pain points and opportunities for future development. [2]

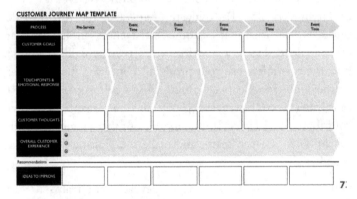

By representing the product flow visually, story or journey mapping also provides a mechanism by which new features and stories can be grouped in a way that, if released together, will preserve the overall integrity of the customer journey. The goal being to create and preserve a pleasurable end-to-end experience for the customer while avoiding the potential of a

---

[2] http://lesliedesigns.com/wp/wp-content/uploads/2014/03/customerjourneymap.jpg

disjointed customer experience. In the reverse, it also allows independent features and stories to be inserted and released knowing that the overall journey will be enhanced without inadvertently creating a negative experience.

For complex architectures with multiple systems, this is also a technique that can be valuable as a view across the entire suite of systems and components required to create and support the customer journey. A customer journey view combined with a high-level architectural view of the systems can be a method by which the business strategy, product, enterprise architecture and development teams can all see clearly what the current state of the product is, as well as what the future state is intended to provide at various points in time. For example, a company that releases monthly might have a current state view plus a monthly view for each of the next three months, effectively capturing the roadmap of both product functionality and architectural improvements targeted to be delivered in each of the next three releases. This can also be extremely valuable as a method to align with defined business goals, outcomes and KPIs because those will define and guide what should be prioritized for inclusion in each of the "future state" snapshots.

The current state of the architecture and the methods and frameworks used by the teams will determine to what extent this is a viable activity. Companies using Scrum with quarterly releases on a multi-layered legacy infrastructure may find benefit in this while a company that is DevOps oriented with full CI/CD on a complete microservices architecture would not benefit from this since the move from "current state" to "future state" is continuous.

While the value stream mapping technique mentioned in the previous chapter seeks for improvement opportunities in getting a product or service to the customers, there is also a need to identify improvement opportunities in how the customers use a product or service once it is available to them. User story mapping can be a very powerful technique not only to capture the customer journey, but also can help people in the organization new to user stories to get a better handle on how to identify and write those stories. Combining user story mapping with value stream mapping can become a way in which to ensure you are building the right things at the right time.

Understanding the customer journey and where it needs to evolve is paramount to effective planning but, as Eisenhower said, planning is essential but plans are useless. There is another appropriate quote that is even more succinct:

> *"Everyone has a plan until they get punched in the mouth"*
> Mike Tyson

You do not know in advance what your customers may or may not do. Whether you are mapping your customer journey and planning around it, or not, you must watch customer behavior as closely as possible. You must see what happens, adapt to customer behavior and momentum and continue to follow to where they lead you. This is why customer behavior, as the outermost feedback loop of "concept to cash" is what encircles the entire organization:

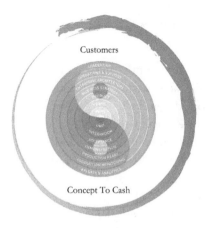

The tighter the concept to cash feedback loop becomes, the more in sync you are with your customers' changing wants and needs. Once you have achieved that state you can begin to watch for synchronicities, being actions, events or changes in customer behavior that stir something in you that calls you to action. When what they need is what you feel is right to do, everything you have put in place to build and learn quickly allows the company and the customers to come together for mutual benefit. When achieving this state, you will look back and realize just how far you have come. The company will be in a completely different state, and you will realize just how much of a better place it has become.

Culture Change Through Collective Mentality

Culture is the collective mentality of the organization. It is the sum total of the personas that individuals take on when they begin their work day and what they leave behind at the end of the day. If the environment in the company is one where employees do not know their value and are misaligned, it pits them against each other resulting in finger pointing, blaming, back stabbing, condemnation, "CYA", and all kinds of self-destructive behavior. That type of environment distracts everyone from the transcendent purpose of the company. What you are left with is a culture of fear.

> *"Fear is the path to the Dark Side.*
> *Fear leads to anger. Anger leads to hate.*
> *Hate leads to suffering"*
> Yoda

True and lasting culture change is the removal of fears and suffering through the environmental changes made as part of the transformation initiatives. Holistic Agility does not lead off with the intent of culture change, but it is a natural result of what the company does for its people along the way.

We cannot possibly help others until after we help ourselves. One of the main analogies in this book has been about a snake shedding its skin at the point where new growth is needed. Employees helping each other to solve small shared problems at their various touchpoints builds relationships and mutual understanding, allows both groups to work together to solve problems and grow, and enables them to help each other and themselves to make improvements. Each time this happens, old skin is shed, and new growth is enabled. The work on smoothing out touchpoints through feedback loops leads to a collective mentality of assistance, learning and improving that will then begin to radiate outward after the inside of the organization has become unified and whole. Bringing disparate groups together at feedback loop touchpoints is a powerful tool because it generates compassion and empathy on both sides as they gain an understanding of each other's pain points. When they spend time together to understand one another they realize they are not in this alone. Everyone is in the same environment and can see through to the same root cause issues from their own perspective and through their own lens.

Seeking out collaborative communications with other groups, understanding them and their struggles, and coming together under a single banner for a common cause builds the momentum necessary to succeed. Employees will realize that they have much more in common than the few things that make each of them uniquely different, although it may not be easy at the beginning if the environment has caused groups to blame one another up until this point.

> *"There is nothing more frightful than ignorance in action"*
> Goethe

If you remain ignorant about "the other" group and their struggles it becomes easy to fall into the trap of blaming them for your collective problems and failures. Sociologically, personal ignorance turns into fear which leads to hate and violence. Luckily in the workplace violence is rare, but the same result masquerades as "throwing people under the bus" type behaviors. As mentioned in the chapter on managing change, looking squarely at that shadow side of ourselves with the intent of transformation is an act of courage. Many people avoid that deep introspection and will lash out at those around them and find anyplace else to lay blame for their pain and discomfort. There is no blame. For anyone around that type of person, it is imperative to practice forgiveness and compassion, if for nothing less than the sake of the collective good being driven toward through the transformation efforts. If that individual is not ready to hear the message, then it is very likely that there is not anything you or the company can do for them at this time. Wish them well and hope they become receptive to it at a later date, but that may have to be somewhere else. For Holistic Agility to continue to gain momentum, each group must take that hard look inward, accept and integrate what they find, then turn back outward to leverage the sensory nature of the organization to fulfill the company's purpose.

The diagrammatic representation of the organization with its concentric circles of groups and feedback loops has been overlaid on the well-known symbol the Tao. That symbol is a holistic representation of oneness with Yin and Yang constantly becoming each other. Translated, Tao means "The Way". There is one last thing that truly brings all of these layers together in a cohesive whole and creates "the way" for Holistic Agility to be realized. Each person in each group in each layer must have a voice and be able to say when something needs to be done, corrected or improved. They need a simple form of direct access and representation into the list of all possible work that the company may do to create value. It creates an infinite loop that exists for the entire lifetime of the company. There is another symbol that represents this final feedback loop that surrounds the entire organization. The ouroboros.

In medieval alchemical tradition, the ouroboros represents introspection and eternal cyclicality. It is the image of a snake swallowing its own tail (see, I can't get away from the snake symbolism even if I tried). It is constantly re-creating itself. This symbol of wholeness and infinity is represented as a type of feedback loop unto itself.

The Holistic Agility equivalent of the ouroboros represents a constant flow from the holistic organization *inside* that is communicating with the environment *outside* and determining the next step one step at a time. Looking behind the concentric circles diagram reveals a never-ending flow of new ideas back into the development teams then back out through the organization, creating even more new learnings and input to be sent back to the development teams, ad infinitum.

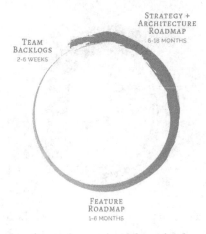

STRATEGY +
ARCHITECTURE
ROADMAP
6-18 MONTHS

TEAM
BACKLOGS
2-6 WEEKS

FEATURE
ROADMAP
1-6 MONTHS

The organizational ouroboros is a funnel where everyone in the organization can insert ideas and suggested work items at any level while it circles back as input to the Teams at the center.

Achieving a meaningful change in culture through Holistic Agility allows the company and its employees to rediscover the power of what is inside. By allowing the outside world to unfold before them and observing and inspecting the details of what is emerging, they are able to use their agility to adapt and evolve and revel in the abundant flow of true benefits for everyone involved.

The Plight of PMs and Middle Management

Throughout this book much has been said about the detrimental effects of a project and plan-driven mentality. There are a lot of project managers, and managers in general, that begin to wonder where they fit in as Agile begins to take hold in their organization. An Agile transformation creates a loss of identity for them. If they are not tracking things like budget, timelines and resource allocations, then what do they do now? It is all too easy for them to get lost in the shuffle, and many people may write them off completely, but they are still valuable to the organization. However, the value they bring comes in a very different form which may be difficult for them to transition to.

Taking a project-by-project view of the work to be done often obscures the holistic view of what teams are expected to accomplish. The various business strategy and product people spin up multiple projects for new functionality while the technology upon which the products are built is also constantly updating and changing through their own set of projects. This is where the holistic view is obscured because all this work ends up being funneled to the same teams, yet when status reports are given it is "project X, Y and Z are on track" all with different timelines, priorities and stakeholders. Each of the various stakeholder groups end up with a false sense of certainty that their projects will be done by the estimated date. That false sense of certainty gets shattered when those expectations are not met because the teams could not complete all of the projects in parallel. That typically results in people complaining about "missed commitments", "do a lessons-learned", and various attempts to lay blame and prevent it from happening in the future, and then it happens all over again the next time around.

If a holistic view of the demands put on the team is made visible, it allows all the various stakeholder groups to understand not only each other's business needs, but exactly just how much they are loading on to the same people that have a limited capacity. There is a chasm between business demands and technical delivery that needs to be bridged. This is where project management professionals can be of enormous value. Through the demands of having to track projects PMs have learned to speak the language of both the business and technology people. By finding a way to bring the two groups together and allowing them to simply spend some time with one another, they will begin to understand one another. A natural compassion and empathy will be generated that can grow into true partnership between the business and technology teams. Such a partnership is what builds that bridge and makes it strong but getting them together in the first place requires something that replaces the standard status reports. They need to be oriented around business outcomes.

It should be noted here, for teams using the Scrum framework, that project managers are not a direct match for the scrum master role. Oftentimes companies implement Scrum and tell their existing PMs that they are now responsible for the scrum master role, perhaps with some basic training before beginning to do their first sprints. These people have had their foundation pulled out from under them and begin to look for familiar footing by tracking when items will be done, even converting daily stand up meetings into de facto status reporting forums. The team members face the "scrum master" to give updates instead of communicating to each other and collaborating on the next day's activities. Scrum masters are champions of the process as well as servant leaders who shine when their team needs their help. For a well-established team that has gone through the growing pains of transformation and addressing their environmental impediments, the scrum master role may not even need to be staffed by a unique individual. I've worked with teams that were mature enough to rotate the role amongst themselves, effectively increasing their capacity for the same overall team/headcount cost. A project manager should not be put unwittingly into a scrum master position if they do not understand the intent is to be a champion and servant leader for the team and be willing to work themselves out of a job.

Project management organizations are also very likely to apply a project-driven mindset to Agile concepts during and after training courses. Many gather scrum and agile-at-scale certifications with the same vigor that they did when going through their Project Management Professional (PMP) training. What may initially feel familiar and comfortable to project managers becomes an impediment to the organization's growth toward Holistic Agility. The big complex frameworks and all the related training and certifications may be familiar and comfortable, but what happens when they return to the heat of battle in the office is that they revert to their deeply ingrained behaviors while using the new terminology. They believe they are "being Agile" when they are not. They simply don't know what they don't know, and that is OK. They need to be placed in a position of value that allows their strengths to be leveraged for the good of the organization.

Project management skills, when used at portfolio and program levels, align quite well with Holistic Agility. Even though the three "Project Management Triangle" variables have been reduced to one after funding teams and establishing a delivery cadence, tracking that last remaining variable of scope against planned KPIs can become the new Red/Yellow/Green report. Tracking the original PM variables is oriented around constrictive and negative aspects of staying under budget and on schedule. There is now the opportunity to change the PM mindset to one of positive expansion and increase of value delivery and process optimization.

By focusing the entire organization on delivering value to customers as quickly as possible, the project management function can shift to track metrics such as planned versus actual values on KPI delivery, time to market (duration from roadmap to production) and ROI at the team, program, portfolio or company levels. These types of metrics will align much more with what will most likely be on something like a balanced scorecard that executives and senior leadership will use to determine overall performance. The challenge for project managers, especially those who have done the job for a long time, will be to stop asking for dates for when things will be done, and start asking how effective were the things that have already been done. When they begin to focus on and drive toward quickly learning KPI impacts, not plan or date driven milestones, it changes the belief of what is important. Having project managers who are tracking against KPIs and scorecard items completely changes the conversation when they are asked "what is the path to get back to Green?"

In the same vein as project managers, technology managers have been trained and conditioned over a very long time. Their job has been to make sure they know exactly what each of their people is working on and to make sure all of their "resources" are fully utilized. As new projects come in, resource leveling through spreadsheets is common, and may be a significant part of the role of a technology people manager. With funded teams that are stable over extended time frames, where do the managers turn their attention? They now must become concerned not with determining what their "resources" are working on but switching their attention to the development and growth of those people.

This is one of the most difficult aspects of any Agile transformation. People who have been drawn to and placed in technology management roles have done so because they were adept at managing the numbers through spreadsheets to get the resulting 100% utilization target they were going for. They often have a very analytical mindset, likely having come up through a technical career path. This frequently comes with more proclivity to introversion and alignment with a "command and control" environment. To take that all away from a technology manager and tell them they are now a "people person" is bound to be met with resistance. This is where a true sense of compassion is required from those leading the transformation initiative because these are human beings we are talking about. In the same way you can no longer expect managers to treat their employees as fungible "resources", a transformation effort cannot demand the same of the management team. They will require support and patience.

> *"He who would learn to fly one day must first learn to stand and walk and run and climb and dance; one cannot fly into flying."*
> Friedrich Nietzsche

For managers that can make this shift, they will become some of the most valued people in the organization because they are constantly focused on improving the people they lead. They become true servant leaders, making the environment and the teams better through helping individuals to grow and reach their own potential. Some managers may not be able to make that shift, and it is perfectly reasonable for them to find somewhere else where they can play to their own strengths. It is an important role of people operations to determine when the evolution of the company has created situations where certain skills are no longer required and either give people new skill sets and roles, or graciously help them to exit from the company. This is the true "fire" of transformation, where something must be consumed allowing for new light to be created.

Managers who become dedicated to the development and growth of their team members will become the people that drive new levels of mastery throughout the organization. Providing opportunities for employees to pursue mastery of their own skill sets, as the third motivator in the Dan Pink video mentioned earlier, completes the trifecta (with "purpose" and "autonomy") created through establishing high performing self-organizing teams who are working on the most valuable things at all times. Therefore, it is important to Holistic Agility to understand the potential that resides within management individuals that can be realized a part of the transformation.

Geographically Distributed Teams

One of the biggest challenges to an Agile transformation is being in an organization where team co-location has not been cultivated, and perhaps has been diametrically opposed in the name of cost savings. Whether it has been through the utilization of offshore or nearshore staffing, or by having a multitude of physical office locations, or by pursuing an extensive work-from-home policy to minimize office space and real estate costs, co-location has not been perceived as a method by which the company can realize operational efficiencies.

When having a geographically distributed workforce without team co-location the loss of effective communication results in more time lost than what is thought to be saved when comparing hourly rates or the cost of office space. More than half of human communication is non-verbal. For any of the multitude of companies that have had their management teams read Patrick Lencioni's "Five Dysfunctions of a Team", they have learned that well-functioning teams are built upon a foundation of effective, open and honest communication. Not being co-located inhibits team building and therefore limits the capability to become high performing. In addition to effective communication of content and meaning, eye contact also builds empathy and rapport amongst team members, leading to much better willingness for team members to assist each other and look after one another. With team ownership of deliverables being a key objective of improvements at the development team level, the more they care for each other the better the results will be.

Beyond simple team communications and dynamics, tried and true techniques such as Gemba walks, where in-person observation of where the work is being done and interacting with the people doing that work, are not possible when the people aren't in the same place as management. It is also much easier for someone to blame and attack someone they have never met face-to-face. I have personally observed that when a workforce is both fully distributed and in a high-pressure environment that they are both quick to attack other people in different groups and are never fully engaged and concentrating on the task at hand. The entire environment is interrupt-driven. During conference calls you can hear them typing out emails and instant messages and attendees frequently have to re-hash what was just said due to others not paying attention because they were either doing something else or being emailed or messaged by someone else. It is also very common for those interruptions to be coming from people that are on the same conference call at the same time. There is a monumental amount of lost productivity in that type of environment that will most likely never be captured in a financial analysis.

For Holistic Agility to be effective, it does not require that every single person be physically located in the same place at all times. What needs to be understood is that skill set distribution does not work but having geographically distributed complete co-located teams *can*. As long as each team has a clear purpose for their existence relative to the whole organization and included in that team are the skills and wherewithal to get that job done, then they can work together in one location while other teams and departments work in other locations.

All teams, whether distributed or co-located, must also allow every team member opportunities to have their voice directly heard. Many companies that leverage offshore or nearshore teams utilize proxies that end up filtering the communication from team members themselves. To gain the most value out of the teams, and to make them as high performing as possible, everyone must be able to participate directly in open and honest communications.

When looking at creating and funding cross-functional teams as part of a transformation companies can give serious consideration to where their skill sets currently lie relative to the co-location objective. Thoughtful use of distributed teams can be another tool to be leveraged in making a healthy and whole organization.

A Healthy and Whole Organization

At one point in my career I was reassigned to a new technology leadership role at my company. As part of that reassignment I inherited a team that was desperately struggling to deliver a new product. The team was running in a strict waterfall fashion with on-site developers and a project manager while the product manager was in another city and the testing team was offshore. The project had already spent over a million dollars and had virtually nothing to show for it. In working in conjunction with the product organization we reassigned the effort to another product manager who was in the same location as the developers and myself, and together we transformed them into a Scrum team that was co-located, cross-functional and funded as a team over a period of time. What we did not know at that point was that the period of time we received funding for was expected to be the last. The company was preparing to pull the plug on the effort and disband the team.

During the transformation of the team we not only co-located them, we also made a strategic hire of a lead developer that had both the technology skills we were looking to use for coding and test automation and who had also worked in a successful Agile environment. We gave the team some basic Agile and Scrum training and were also able to coach them along the way as needed. The dynamic of the initiative started to shift immediately as the people came together, established a new way of working and focused on delivering value early and often. Within two months the team began to deliver the contents of each sprint into a production environment and moved to weekly releases shortly thereafter, also utilizing A/B testing and other analytics to determine what was being used by the initial customers and what did not warrant further time and investment. The team became a success story in the organization that spawned many more broad transformation initiatives, but their business results are not what was the cause of their success. They were successful because they were *happy*.

By creating a co-located team with a clear purpose and the skills and tools to fulfill that purpose, they understood that they were valuable to the organization and to each other. As little successes began to pile up, they became enthusiastic about what they were doing. They enjoyed coming to work when they began to feel like a part of a team that had purpose and value. They enjoyed it every day. It became the "culture shift" other companies try to manufacture artificially. It was not about having a foosball table or kegerator in the break room. It was about being happy, and making others happy, all as a prerequisite for success.

As this was coming together you could see the epiphanies happening within each one of them. Each time a proverbial light bulb of enlightenment went on and their eyes lit up, you knew something special was happening. They were smiling, energetic and engaged in what they were doing. Creating a happy team was what led to the success they achieved. It was not the other way around. This team was a case in point regarding the creation of the "win-win-win" environment mentioned earlier, but those wins now represent what Holistic Agility is about:

- Happy – A happy and enthusiastic team that has purpose and value

- Healthy – Customers are receiving a quality product from that team that consistently meets their needs in a timely fashion
- Whole – Business success from learning about and meeting customers' needs is being achieved and shared throughout the entire organization

Any healthy and whole organization has these tell-tale signs. The people are happy to be there, and it shows through a sense of harmony across the company. Each group is working together to create a fluid and cohesive company. As the outermost feedback loop of "concept to cash" becomes tighter and tighter, the rest of the organization within begins to hum with an ever-increasing vibration. That higher vibration is a true parallel to that of an individual moving to higher levels of human consciousness. It is awakening to the realization that your perceived rationality has failed you up to this point. Your attempts to find certainty and exert control over the universe of all possibilities has consistently fallen short of what you know can be accomplished. It is only by looking inward and courageously confronting the less than desirable aspects, then returning outward with an understanding what your purpose is that new levels of growth can be achieved. When you allow your purpose to proceed in evolutionary harmony with the constant ebb and flow of the world around you, you come to realize that you have reclaimed an innate power. It was always there, but what you had been taught along the way through education and experience, things that were falling unfiltered into your subconscious mind, became deeply held beliefs that were off the mark. These beliefs were shared with many others, creating a collective set of beliefs and a culture that ended up being very hard to overcome. Once you allow those beliefs to be broken down, and begin to build upon a new foundation, the power you had all along becomes unlocked and unleashed.

For many people it may be a challenge to equate a company's maturity to that of human consciousness, but that is the main foundation to many emerging management and leadership theories, a great example of which is the concept of Teal organizations. Teal Organizations are companies that have evolved through other stages of development to the point of having a life and a sense of direction of their own. Instead of trying to predict and control the future, members of the organization are invited to listen in and understand what the organization wants to become, what purpose it wants to serve. The metaphor on the Teal for Teal wiki says that founders of Teal organizations refer to themselves as living organisms or living systems and goes further to state:

> *"Life, in all its evolutionary wisdom, manages ecosystems of unfathomable beauty, ever evolving toward more wholeness, complexity, and consciousness. Change in nature happens everywhere, all the time, in a self-organizing urge that comes from every cell and every organism, with no need for central command and control to give orders or pull the levers"*[3]

Companies have been and continue to find parallels to things like Maslow's theory of Self-Actualization, such as Teal organizations do, and to Jung's theory of Individuation where experiences over time become integrated into a well-functioning whole while utilizing the "unconscious" (the collective seemingly chance events that transmit discernable meaning) to achieve that goal of a well-functioning, integrated whole organization. Patrick Lencioni has another book titled

---

[3] *https://reinventingorganizationswiki.com/Teal_Organizations*

"The Advantage" detailing the concept that organizational health surpasses all other disciplines in business as the greatest opportunity for improvement and competitive advantage. These all speak to the power of creating healthy and whole organizations.

At this point in time, too many companies are not healthy and whole. They do not have unified objectives and purpose. They don't have shared and streamlined communications. The employees don't know each other as people. They are not happy. Everything contained in this book has been based on the fact that we are all equally human people. People with a common purpose come together to achieve great things. If your organization is characterized by discord and the need for "accountability", realize it is time for transformational change. Your message of transformation needs to be heard, for your sake and for everyone around you, your colleagues, and your customers. The world is changing. If you cannot grow from where you are today, find someplace else where you can. If you can make others around you happy through transformation, do it. There is no time like the present. When you make others around you happier, you'll be happier, your company will reap the rewards, and the world will be a better place for it. Everybody wins.

Only the simplest of organisms can survive when a part of it is taken away and left behind. Organizations pursuing an Agile transformation are not simple. They are complex and varied with a multitude of unique individuals. If an attempt at transformation does not take the whole company into consideration, there is a very high probability that it will become another story of failed attempts, wasted investments and missed opportunities. Change for the better must be brought to everyone involved. "United we stand, divided we fall". Using Holistic Agility as a mindset and method to engage the entire organization, taking every person into account and helping them thrive by playing to their unique strengths, will create more shining examples of companies that have shed their proverbial skin and become truly transformed.

Glossary of Terms

6+6 Budget – A form of a rolling forecast where the first 6 months contain actual expenditures while the second six months contain an adjusted forecast in response to the actual expenditures

A/B Testing – Comparing two versions of a solution to see which one performs better

Acceptance Test Driven Development (ATDD) – A practice in which a team collaboratively discusses and defines acceptance criteria before development begins

Agile In Name Only (AINO) – The use of Agile vocabulary without corresponding behavioral changes

Balanced Scorecard – A strategy management tool that utilizes a small number of financial and non-financial metrics to track and assess organizational performance

Big Data – Extremely large data sets that may be analyzed computationally to reveal patterns, trends and associations especially relating to human behavior and interactions

Capital Expenditure (CapEx) – Money spent by a business or organization on acquiring or maintaining fixed assets

Center of Excellence (CoE) – A team that provides leadership, best practices, research support and/or training for a focus area

Chief Information Officer (CIO) – A job title commonly given to the most senior executive in an enterprise responsible for the information technology and computer systems that support enterprise goals

Community of Practice (CoP) – A group of people who share a concern or passion for something they do, and learn how to do it better as the interact regularly

Continuous Integration (CI) – A development practice that requires developers to integrate code to a shared repository several times a day

Continuous Delivery (CD) – A software engineering approach in which teams produce software in short cycles, ensuring that the software can be reliably released at any time

Customer Experience (CX) – The entirety of interactions a customer has with a company and its products

Definition of Done (DoD) – A set of rules or criteria that a team adopts as a guide for when they can legitimately claim that a given unit of work is complete and ready for review, acceptance and approval

Definition of Ready (DoR) – A set of rules or criteria that a team adopts as a guide for when a given unit of work is deemed immediately actionable and ready to be produced and completed

DevOps – A software engineering culture and practice that aims at unifying software development and software operations

Employee Stock Ownership Plan (ESOP) – A qualified defined-contribution employee benefit plan designed to invest in the stock of the sponsoring employer

Extreme Programming (XP) – A software development methodology intended to improve software quality and responsiveness to changing requirements

Fungible Resources – Goods or commodities whose individual units are essentially interchangeable

Gemba Walk – An informal tour of where the work is taking place and interviewing process participants to gain a comprehensive understanding of the process

Kanban – Japanese for "signboard" or "billboard", it is a manufacturing method that regulates the flow of work through the use of an instruction card sent along the production process

Key Performance Indicator (KPI) – A measurable value that demonstrates how effectively a company is achieving key business objectives

Memetics – the study of information and culture based on an analogy with Darwinian Evolution, showing that the best solutions emerge through trial and error.

Minimum Viable Product (MVP) – A development technique in which a new product is developed with sufficient features to satisfy early adopters

Operating Expense (OpEx) – Expenditures that a business incurs to engage in any activities not directly associated to the production of goods or services

Organizational Change Management (OCM) – A framework for managing the effect of new business processes, changes in organizational structure or cultural changes within an enterprise

Quality Assurance (QA) – The maintenance of a desired level of quality in a product especially by means of attention to every stage of the process of delivery or production

Scrum – Named after an action in a rugby match, it is an agile software development technique for small teams using timeboxed iterations who meet daily to track progress and re-plan

Scrum of Scrums (SoS or S²) – A technique used to scale scrum to large teams where ambassadors of each team scrum participate in a daily meeting with other ambassadors

Release Train – A long-lived self-organizing collection of agile teams that plans, commits and executes together

Return On Investment (ROI) – The amount of return on an investment relative to its cost

User Experience (UX) – The overall experience of a person using a product such as a website or application, especially in terms of how easy or pleasing it is to use